Virginia Woolf

Edited and Introduced by

Rachel Bowlby

LONGMAN
LONDON AND NEW YORK

Longman Group UK Limited,
Longman House, Burnt Mill,
Harlow, Essex CM20 2JE, England
and Associated Companies throughout the world.

Published in the United States of America
by Longman Publishing, New York

First published 1992

British Library Cataloguing-in-Publication Data

A catalogue record for this book is
available from the British Library

ISBN 0 582 06152 0 CSD
ISBN 0 582 06151 2 PPR

Library of Congress Cataloging-in-Publication Data

Virginia Woolf / edited and introduced by Rachel Bowlby.
        p.   cm. – (Longman critical readers)
    Includes bibliographical references and index.
    ISBN 0-582-06152-0 – ISBN 0-582-06151-2 (pbk.)
    1. Woolf, Virginia, 1882–1941 – Criticism and interpretation.
  I. Bowlby, Rachel, 1957–   .   II. Series.
  PR6045.072Z8918   1992
823'.912 – dc20   91–39684
                                                             CIP

Set 9K in 9/11.5 Palatino
Produced by Longman Singapore Publishers (Pte) Ltd.
Printed in Singapore

Contents

General Editors' Preface

The outlines of contemporary critical theory are now often taught as a standard feature of a degree in literary studies. The development of particular theories has seen a thorough transformation of literary criticism. For example, Marxist and Foucauldian theories have revolutionised Shakespeare studies, and 'deconstruction' has led to a complete reassessment of Romantic poetry. Feminist criticism has left scarcely any period of literature unaffected by its searching critiques. Teachers of literary studies can no longer fall back on a standardised, received, methodology.

Lectures and teachers are now urgently looking for guidance in a rapidly changing critical environment. They need help in understanding the latest revisions in literary theory, and especially in grasping the practical effects of the new theories in the form of theoretically sensitised new readings. A number of volumes in the series anthologise important essays on particular theories. However, in order to grasp the full implications and possible uses of particular theories it is essential to see them put to work. This series provides substantial volumes of new readings, presented in an accessible form and with a significant amount of editorial guidance.

Each volume includes a substantial introduction which explores the theoretical issues and conflicts embodied in the essays selected and locates areas of disagreement between positions. The pluralism of theories has to be put on the agenda of literary studies. We can no longer pretend that we all tacitly accept the same practices in literary studies. Neither is a *laissez-faire* attitude any longer tenable. Literature departments need to go beyond the mere toleration of theoretical differences: it is not enough merely to agree to differ; they need actually to 'stage' the differences openly. The volumes in this series all attempt to dramatise the differences, not necessarily with a view to resolving them but in order to foreground the choices presented by different theories or to argue for a particular route through the impasses the differences present.

The theory 'revolution' has had real effects. It has loosened the grip of traditional empiricist and romantic assumptions about language and literature. It is not always clear what is being proposed as the new agenda for literary studies, and indeed the very notion of 'literature' is questioned by the post-structuralist strain in theory. However, the uncertainties and obscurities of contemporary theories appear much less worrying when we see what the best critics have been able to do with them in practice. This series aims to disseminate the best of recent criticism and to show that it is possible to re-read the canonical texts of literature in new and challenging ways.

RAMAN SELDEN AND STAN SMITH

The Publishers and fellow Series Editor regret to record that Raman Selden died after a short illness in May 1991 at the age of fifty-three. Ray Selden was a fine scholar and a lovely man. All those he has worked with will remember him with much affection and respect.

Acknowledgements

We are grateful to the following for permission to reproduce copyright
material:
Editions les Femmes for a translation of an extract from *Virginia Woolf,
Vers la maison de lumière* by Françoise Defromont; the author, A.A.H.
Inglis for his essay 'Virginia Woolf and English Criticism'; the author,
Mary Jacobus, for her essay '"The Third Stroke": Reading Woolf with
Freud'; Novel Corporation for the essay 'Penelope at Work: Interruptions
in *A Room of One's Own*' by Peggy Kamuf from the journal *Novel: A Forum
on Fiction* 16, No 1 (Fall 1982) copyright Novel Corp., 1982; Overseas
Publishers Association and the author, Susan M. Squier, for her article
'Tradition and Revision in Woolf's *Orlando:* Defoe & "the Jessamy
Brides"' from *Women's Studies* Vol 12, No 2 (1986), copyright © Gordon
& Breach Science Publishers Inc; Princeton University Press for the
chapter 'The Brown Stocking' from *Mimesis: The Representation of Reality in
Western Literature* by Erich Auerbach, translated by Willard Trask,
copyright 1953 Princeton University Press, © 1981 renewed by Princeton
University Press; Routledge, a division of Routledge, Chapman & Hall
Ltd, for the chapter 'The Island and the Aeroplane' by Gillian Beer from
Nation & Narration edited by H. Bhabha (pub. 1990); Routledge, an
imprint of Routledge, Chapman & Hall Inc, for an extract from the essay
'Woolf's Room, Our Project: The Building of Feminist Criticism' by
Catharine Stimpson from *The Future of Literary Theory* edited by Ralph
Cohen (pub. 1989); University of New England Press for the article
'Narrative Structure(s) and Female Development: The Case of *Mrs
Dalloway*' by Elizabeth Abel from *The Voyage In: Fictions of Female
Development* edited by Elizabeth Abel, Marianne Hirsch and Elizabeth
Langland, copyright © 1983 by Trustees of Dartmouth College.

A Note on Editions

In 1992, for the reason explained in the Introduction, various new
editions of Woolf's works appear. Since those used by the contributors to
this volume are likely to be superseded, it did not seem worthwhile to
standardise the references to a single edition.

1 Introduction

In 1992, the year that this collection is published, Virginia Woolf, who died in 1941, has been dead for fifty complete years. This simple fact will have all sorts of effects, measurable and unmeasurable, ascertainable or barely to be guessed. Something is bound to happen to her, because this is the moment when legally, in Britain, she will make a forced and highly visible movement: she will 'enter the public domain', or (to use a phrase which looks like much the same movement viewed from the other direction) she will 'come out of copyright'. Virginia Woolf will do this whether she likes it or not, more dead and more alive than she has ever been. Henceforth, any publisher may publish her published works (at least, those published in Britain and before her death) without having first to seek permission from the single publisher which up till then held the sole rights to her publication (by the end it was Chatto and Windus; for most of the time it had been the Hogarth Press founded by Leonard and Virginia Woolf themselves).

What will this entry, or emergence, look like? It will no doubt (I write in 1991) be seen and remarked upon with as much evidence and noise of various regulated kinds as though some great lady, a royal personage perhaps, had descended the stairs from her private room to process with some ceremony down Whitehall. For Virginia Woolf is not just any writer, not just some second-hand out-of-print author whose belated sortie from copyright is as non an event as every other year has been since their deaths in print and in life – and the fact that you may have seen in the royal person and the Whitehall walk some familiar textual phantoms from *Mrs Dalloway* and *A Room of One's Own* is one indication of this.

In Woolf's case, the occasion has long been planned, and by the time of your reading this there will be several more paperback editions jostling for priority on bookshop shelves with the probably by now distinctly fading productions of Harcourt Brace or Granada, depending on whether we are imagining ourselves to be in Virginia or Wolverhampton. The new books will come furnished with all the canonical paraphernalia

1

which they have up until now been denied or spared, in the form of editors' introductions and footnotes at the back, with masses of useful material for students and teachers on the many courses for which Woolf is a prescribed author, and also for the many less readily classifiable individuals who may fall into the category of what Woolf, following Samuel Johnson, called the 'common reader'.

The fifty-year anniversary focuses one kind of mark in the history of how an author is read, by how many and in what contexts. The appearance of all these new paperbacks will be one more occasion for numerous reviews – in academic journals probably less than in the book sections of the 'serious' newspapers and weeklies. And this part of the event will be just as significant an element in the manifestation of Woolf's change of legal status as the books themselves.

As one of the 'Bloomsbury set' whose lives and personalities seem to provide a source of endless fascination in the form of biographies and editions of letters and diaries susceptible to being picked up and extracted in the papers, Woolf is already well in line for notice. Her intriguing life and close involvement with others equally extraordinary, all reflected more or less directly in her writings both private and public – in her diaries and letters and in her novels and essays – make her a key representative of a certain kind of literary and sexual life in the first half of this century.

On the other hand, as the subject of an ever-increasing quantity of books and articles about her novels, Woolf is less in line for these highly public spaces, which do not make a habit of paying column-inch attention to a type of academic writing which is considered far too specialised for the general reader's interests. There is this Woolf who features in academic criticism and is often regarded quite simply – though at the same time anything but simply, for this occurs in numerous modes, not all of which are mutually compatible – as the most important woman writer of the twentieth century. And there is the other Woolf – or more usually the Virginia Woolf – who features in more widely available writing and other media productions.

In this connection too, Virginia Woolf will be entering a new space – or rather, one Virginia Woolf not well known outside her normal institutional quarters will be taking a turn alongside, and perhaps in some uneasy conflict with, a better-publicised sister known for quite different qualities. For the introductions in the new editions, written for the most part by academics, but written in a style that is meant to be comprehensible to readers with no particular training in contemporary literary criticism, are likely to carry through enough of the current preoccupations in Woolf criticism of the more esoteric kind for these in themselves to become matter for comment.

Of course, these two Woolfs are not as separate as I am making them appear for the purpose of setting up some debatable oppositions between the academy and the media, the restricted and the public, or (to return to the starting point when Woolf appeared to be on her way out into a new space) between the inside and the outside. For one thing, the 'restricted' side of the division which I have so far associated with the academic looks quite the opposite from within that enclosure, in so far as it is one. First of all, academic critics tend to look out, and sometimes down, on the more public media as the locus of limits on what can be said, in relation to which their own writing is perceived as relatively more free from circumscriptions. And secondly, following from this, the academic treatment of Woolf has often used her, or her writing, as the testing ground for questions and speculations about all sorts of topics of current concern, both academically and elsewhere.

Woolf has been used as a way to think generally about women and writing, or women and language. But she has also been used to discuss women and psychology, women and sexuality, women and mothering, masculinity and war, women and national identity – to name just a few of many overlapping areas. All these are topics that Woolf herself addresses more or less extensively in her own work; what is significant, however, is the way in which this can function as an automatic point of reference in so many different fields which might seem at first sight to have little to do with a body of work which started critical life as no more – and no less – than that of a modern English novelist.

So the terms of the division are anything but clear, and vary according to the side from which one is looking. But there are also some books which do not fall simply into one or other category, as exclusively either academic or more popular, but not both. A book like the one by Louise DeSalvo on Woolf's sexual abuse as a child, which appeared in 1989, is a case in point.[1] Although it was written by someone known as a scholarly critic of Woolf, it was marketed and duly reviewed in such a way as to straddle both fields, treating a topic of contemporary urgency and highlighting an aspect of Woolf's autobiographical writing which had received little attention until then. DeSalvo's book, like others which isolate a single part of Woolf's life and writing raises questions about how different frames or contexts for understanding particular writers, and especially this particular writer, come into prominence.

This interest in contexts – how to define them, how to think about the limits they set and the possibilities they suggest for thinking – is itself an example of a contemporary context, and this introduction will be exploring it further in relation to the history of Woolf criticism. But before we begin to look into that, it is worth noting too that distinctions between controlled and free spaces of writing and thinking, between the insides and outsides of institutions, as well as between the private insides and public exteriors of men and women, are themselves objects of Woolf's constant reflection, as she repeatedly sets them up and then challenges both their usefulness and their validity. In *A Room of One's Own*, most memorably and most famously, Woolf's imaginary female narrator is forever being turfed off restricted grass or kept out of restricted libraries in the masculine place called 'Oxbridge'. But this situation is seen as being advantageous as well as more obviously not, since the woman is thereby freed from the conformities and conventions which in one sense make her male counterparts more the prisoners than her who neither has nor is offered recognisable forms of social and institutional identification.

Elsewhere, and especially in her numerous essays on critics and criticism, Woolf frequently addresses questions that have to do with its styles and provenance. In what schools of education or literature was the writer formed? What kind of a reader does s/he imagine himself – it is usually a man – to be writing for? She is also fascinated by the history of criticism and how this is related to the various hypothetical readerships for which it was intended at different times, and its various degrees of accessibility, in terms of both market and style. In one way, she appears as a strident anti-critic in her often asserted belief in the desirability that each reader should form her own conclusions, without the needless complications or distractions of others' interpretations. But in another way, her own practice belies this constantly, for her essays are a model of suggestive and idiosyncratic reading of a kind which is clearly meant to inflect her own readers' readings of the author in question.

It follows from this that from one point of view, Woolf might seem to be someone who could be called upon as an antagonist practically *avant la lettre* of the proliferation and professionalisation of literary criticism that has been such a striking feature of twentieth-century publication. But the kind of critical writing she does suggests that she might also be regarded as an advocate of the view that there neither should nor need be an absolute distinction between the categories of literature and criticism. Woolf's critical essays tend to read more like stories than articles, and all the more for their posture of emanating from the naive impressionability of an ordinary, untrained reader. She wrote quite consciously as someone with no formal education in the matter; but even here the oppositions do not function in any simple way, since at the same time she also wrote most of her essays in a strictly professional way: to deadlines, at set lengths, and primarily to make money.[2]

Woolf's own practices and postulations in relation to criticism form an appropriately complex backcloth to the criticism written about Woolf herself – to that amorphous collection of texts which constitute what is curiously known as the history of an author's 'reception', as though to signify her passing over into a semi-public, relatively hospitable domain, a kind of official party held in the best room of a nonetheless private residence. And it becomes obvious straightaway, from the perspective of the contemporary questions which we are prompted to ask, that Woolf criticism exemplifies many of the tensions, fruitful and unfruitful, of debates both within Woolf's writing and subsequent to it, about the proper or possible functions or aspirations of this genre.

Erich Auerbach's *Mimesis*, the book whose final chapter begins this volume, was written during the war and translated immediately afterwards (in 1946).[3] It became a classic text not only for the developing discipline of comparative literature, but also on the many introductory and survey courses of Western literature that were set up as part of the postwar expansion of higher education in the United States and elsewhere. Auerbach's idiosyncratic technique is to take sample passages in texts from successive periods and different literatures: he begins with a famous chapter on Odysseus's scar, and deals along the way with works that include Genesis, the *Chanson de Roland*, Montaigne's *Essais* and Emile Zola's *Germinie Lacerteux*. The book is both a distinctive history of some of the different Western forms of what the subtitle calls 'the representation of reality', and also a series of demonstrations of what can be elicited from the close reading of brief extracts.

This is true of the chapter on Woolf's *To the Lighthouse* as of all the others; but there is something more too, for in this concluding case, Auerbach uses Woolf, his representative of modern writing, to make a retrospective point about the methods of his entire study:

> I see the possibility of success and profit in a method which consists in letting myself be guided by a few motifs which I have worked out gradually and without a specific purpose . . . for I am convinced that these basic motifs in the history of the representation of reality – provided I have seen them correctly – must be demonstrable in any random realistic text. But to return to those modern writers who prefer the exploitation of random everyday events, contained within a few hours and days, to the complete and chronological representation of a total exterior continuum – they too (more or less consciously) are guided by the consideration that it is a hopeless venture to try to be really complete within the total exterior continuum and yet to make what is essential stand out.
>
> (541)

What this brings out is the largeness of the claim that is implicitly being made for modernist writing and, in particular, for Woolf. The focus on 'random' details and fragments, which Auerbach has analysed through the episode at the start of *To the Lighthouse* when Mrs Ramsay is measuring the stocking against her son James's leg, becomes nothing less than the organising principle – yet an organising principle which deliberately sets itself against the obvious organisational outlines of major 'exterior' events – for the whole understanding of the history not only of Western literary realism but, by extension, of Western ways of conceptualising reality.

One upshot of this is Auerbach's linking of critical to literary writing in such a way that one cannot be seen as independent of the other. Criticism is not some adjunct or afterthought that follows after its object in an ancillary, secondary and qualitatively different mode. Its function is in part to draw out the tendencies and potentials of literary writing in relation to larger, not strictly literary concerns. We shall see this operating with other critics in relation to recent criticism of Woolf; in Auerbach's case, the implications are no less than global in their import. What he sees in the modernist writing epitomised by Woolf's novel is not only a truer approximation to the experience of life, but also the intimation of something which may provide a remedy for the conflictual predicament of the world in the aftermath of fascism and the war: 'It is precisely the random moment which is comparatively independent of the controversial and unstable orders over which men fight and despair; it passes unaffected by them, as daily life' (44). Auerbach's dream of an 'approaching unification and simplification' may seem too simple forty years on; but what is also striking now in his vision is its deployment of a modernist aesthetic represented by Woolf as a contributory and anticipatory element in the making of a better future.

The position being granted to Woolf here should not be underestimated. This is the first time that she has been accorded the prime position among her fellow modernists, let alone placed at the culmination – and indeed as the avant-garde for a new order – of the entire Western tradition of literature. At this time, although the establishment of an order of merit among modern novelists who wrote in English was by no means either fixed or regarded as a necessary task, Woolf was not habitually mentioned as the author of masterpieces which had changed the conditions of literature, if not of the representation of reality in general. Modernist writing was itself far from being taken as a matter for automatic appreciation or canonisation. But in contexts where it was taken seriously, Woolf's name, though mentioned with the respect due to one of the leading 'experimental' novelists of the interwar period, did not figure at the head.[4]

The high reputation which is hers today was not, however, established as a result of Auerbach's promotion of her significance, despite the wide influence of his book, and the regular references to and reprintings of 'The Brown Stocking' as a model for the possibilities of a certain kind of reading. Tony Inglis's piece, written in 1974 and published for the first time in its original English in this volume, offers a different example of the uneven and unpredictable developments of Woolf criticism. Its own case for Woolf's importance was effectively knocked into silence before it had been given a hearing because of the momentous transformation of Woolf and her criticism that had been completed a few years afterwards, and which we will come to in a moment.

Inglis's essay was written for an important French colloquium on Woolf and Bloomsbury; its proceedings were published in French in a widely available paperback edition, but never translated into English.[5] The occurrence of such a conference at all testifies to the French interest in Woolf – centring at the time on Jean Guiguet, the author of a major study of her work in the 1960s – whose continuation in feminist and other forms can be seen in this volume from the extract of Françoise Defromont's book. Inglis's contribution was intended, as comments by himself and others make clear, to be an attack in the style of F. R. Leavis and his followers on the Bloomsburyesque Virginia Woolf they regarded, by reputation and sometimes in print, as of little literary or cultural value.[6] Inglis turns the tables on this interpellation in a number of ways, one of which is to show, through D. W. Harding's reading of Shelley, that Leavisites had not always proved unable to admire literary qualities comparable to Woolf's. But much more significantly, he makes a double argument in relation to the future directions of Woolf criticism, in the context of her relatively low valuation at the present time. He suggests that Woolf's own advocates have been too parochial in their failure to make their claims for the importance of her writing outside their own limited domain. And he sketches that possible case himself throughout his own argument for the need for a revaluation of modernist writing in Britain, in which Woolf would occupy a much more central place than previously.

Inglis's attempt to shift the grounds of Woolf criticism from English moralism to European modernity, philosophical and literary, was not followed at the time, even though the direction taken by English studies in Britain and the United States not long afterwards, towards European developments shorthanded in the word 'theory', ought to have fostered this. Many of the critical studies of Woolf in recent years have indeed been every bit as theoretical, in this 'Continental' sense, as in other areas of English, but this has occurred as part of another change which has altered their appearance beyond recognition. Since its beginning in the early 1970s and throughout its dramatically rapid development ever since, Virginia Woolf has occupied a central place in the new field of feminist criticism.

In this context – so obvious a context that it has become difficult even to see it as something which has not always been – the question of Woolf's standing in relation to other early twentieth-century novelists is not so much absent as displaced by what now seem far more pressing and exciting issues to do with the fact that she was a woman. Feminist criticism has transformed the terms on which questions are asked in almost every sector of literary criticism in general, and has drawn on and argued with all the other new methodological fields – structuralism, deconstruction, psychoanalysis, to name the most obvious – associated with the general turn to theory.

The output of articles and monographs on Woolf is now vast: just how many there are I discovered when I ordered a computerised printout of the published material from the last four years and found that the number of items ran, or sprinted, to past the five hundred mark. In such a context, the notion of constructing something that could pass for a representative selection or a choice of the best very quickly comes to look like an impossible task or dream. This is all the more true because Woolf criticism has reached out to touch upon so many different feminist areas. Almost all of it relates to Woolf as a woman writer, or (and the two most often overlap) to feminist questions of all kinds for which Woolf's writing is taken – as with the case of child abuse – to be a suitable reference or testing ground.

Already by the time that Tony Inglis's lecture was delivered in 1974, when feminist criticism barely had a name, and when what there was had been directed more to rereadings of male writers than female writers, there had appeared a number of feminist books wholly or partly on Woolf, predominantly dealing with her in connection with the idea of androgyny.[7] By the time of Toril Moi's *Sexual/Textual Politics*, published in the mid-1980s, it could seem natural to begin a general work on feminist criticism and its relation to theories of female subjectivity with a chapter on Woolf, who by now figured widely as something like *the* woman writer in person. This is one reason why writing on Woolf is a less specialised province of criticism than writing about other women novelists and poets. Almost every feminist critic has had her bit to say about Woolf, and many of those included in this anthology are primarily known for their work in other periods or on other authors. But there is also plenty of specialisation in feminist Woolf criticism; much of it in the United States is associated with the work of Jane Marcus, who has made a particularly strong case for seeing Woolf as a political and sexual radical.[8]

After the pieces by Auerbach and Inglis, then, it will come as no surprise now that the rest of the book is entirely written by women. All the subsequent chapters take it for granted that Woolf's writing cannot be considered apart from feminist questions which range from the specifically literary to those that are global in the same way that Auerbach's concluding claims turned out to be: they take Woolf's writing as a means of looking at general issues of current concern to feminism. But by the same token, the articles by no means follow a smooth course of development, either chronologically or in relation to one another.

Precisely because Woolf has come to be a focal point for feminist arguments, literary and otherwise, there is no consensus about her writing. At times, indeed, it seems as if Woolf can be all things to all feminists, so different and often contradictory are the interpretations of what her texts mean in relation to contemporary questions. Any collection of recent writing about Woolf is going to have something of the look of an introduction to the range of recent feminist thinking too (see Mary Eagleton's *Feminist Literary Criticism* in this series, Longman, 1991).

But this diversity and divergence is also, I think, an effect generated by what – to cite one of a number of phrases from *A Room of One's Own* which have both passed into the everyday currency of feminist criticism and, at the same time, become miniature texts for interpretation – we might call the 'difference of view' within Woolf's own writing, which often changes direction many times within a single text. To me, this quality of contradictoriness is not something to be reproached, but rather a sign of Woolf's writing being genuinely experimental: not only new in its styles and movements, but rough and unready, not quite slipping into a form or system which could be identified as fully finished, all sewn up and complete. Woolf does not hide her uncertainties. Asking a question does not have to be followed by something that will appear to have answered it once and for all; in fact it is rather as though such an appearance of finality, of coming to a conclusion, would represent more the death than the highest achievement of thinking.

The fact that these kinds of arguments can rage and love over someone who never regarded herself primarily as a theorist of feminism or of anything else – 'Theories . . . are dangerous things', she remarks in one of her essays[9] – says much about what has happened to the scope and (in a positive sense) the pretensions of the kind of writing that is still produced under the increasingly unlikely name of English literary criticism. It puts Woolf unexpectedly in the same situation, in this respect, as a contemporary thinker in an area that appears to have little in common with that of a novelist. But Freud, like Woolf, has been the object of heated as well as coolly academic debates which regularly turn upon issues that have to do with his writings not having the appearance of a perfectly consistent finished article, with no loose ends or open questions, no tantalisingly visible points of contradiction. And perhaps this is partly what has brought Freud and Woolf so surprisingly together so often in recent years: they share this unpredictable unevenness which makes it impossible to reconstruct a smooth and faultless developmental path for their practice and thought, and which also makes them so full of future possibilities.

There are other, more evident reasons, of course, for the Freud–Woolf connections especially apparent here in the pieces by Elizabeth Abel, Françoise Defromont and Mary Jacobus. From the beginning, feminist literary criticism has been interested in questions of male and female psychology, looking at texts with an eye to what they may indirectly be suggesting about the differences or similarities of the sexes. As I mentioned above, there was a period in the 1960s and early 1970s when Woolf's writing was taken up in support of a notion of psychological androgyny, according to which characters – and, by extension, actual people – are not simply confined to the qualities normally associated with one sex, but show various mixes and combinations of masculinity and femininity.[10] This approach tends to point optimistically towards the possibility of a growing harmonisation of the personalities of both men and women, with the recognition that feminine and masculine aspects should be seen as being of equal and compatible value.

The androgyny model was hopeful in that – in this respect like Auerbach – it looked towards a potential simplification and unification of minds that were only contingently troubled and divided. Since the mid-1970s, however, the emphasis has shifted towards an exploration of difficulties and irresolutions seen now to be inherent in human subjectivity, and particularly female subjectivity. In this context, Freud, who was previously dismissed by feminists as a misogynist of no interest other than as a bad, normative influence on therapeutic practices who needed to be debunked, has entered the picture in a different light. Following the publication of Juliet Mitchell's *Psychoanalysis and Feminism* in 1973, the possible uses, if not the indispensability, of psychoanalysis for understanding female subjectivity have been a subject of constant debate.

This interest in psychoanalysis applies not just to literary criticism but to feminist theory in general. Yet it is striking that many of the leading contributors to this debate in Britain and the United States have themselves been women whose background was in literature (Mitchell is the first example). More recently, the adoption of questions derived from versions of psychoanalysis for feminist readings of literary texts by men and women has been a major factor in revitalising the hitherto somewhat stagnant area of 'Freudian' literary criticism. The work of the French psychoanalyst Jacques Lacan has been one major influence, implicitly manifest here in Defromont's and Jacobus's pieces. Another, rather different point of reference, discernible in Elizabeth Abel's study of Woolf, is the work of the American Nancy Chodorow, whose *The Reproduction of Mothering* (1978) has provoked take-ups and counter-clarifications in many fields of feminist theory in the United States, and has been much utilised in studies of the closeness of the mother–daughter bond lying behind later formations of relationships between women.

The extract from Françoise Defromont's book translated here is just prior to the part where she goes on to talk about questions of sexual difference, and in particular how a woman's wish to write, or her difficulty in doing so, may involve a psychical negotiation of masculine identifications which cannot be simply referred to a sociological bar. In this section, moving between passages from several of the novels, Defromont leads into the complexities of these questions of sexual identity by looking at pivotal scenes of fragmentation, violence and sexuality which seem to underpin and undermine – both at once – the structures of Woolf's novels, as well as the narratives of her own life.[11] In undertaking symptomatic readings of both the literary and the personal writings (diaries, letters, autobiographical fragments), Defromont's work is part of a new interest in taking up once again the question of what might be said about the relation of an author's biography and her work, without seeing the first as simply the source or determination of the second. The psychoanalytical postulate that a life is not so much a given ground of truth from which fictions of literary and other kinds have a measurable difference, but itself made up of stories, both personal and cultural, told by and to and for the subject positioned as an 'I', as first-person narrator, has proved to be very fruitful.

Abel's piece emphasises the relations between Woolf's novels and Freud's psychoanalysis in terms of their comparable constructions of narratives of psychic development which impinge in incompatible ways upon boys and girls. In the 1920s, when Woolf wrote *Mrs Dalloway*, Freud was rethinking what he had previously assumed to be an approximate parallelism in the typical stages of how children of either sex move from the biological maleness or femaleness given by nature to the social masculinity or femininity ordained by culture. He came to see that the cultural dominance of masculinity implied that there was no symmetry in the paths taken by, or imposed on, boys and girls in their movement towards the sexual identities of adulthood. Girls, with nothing to lose, can only learn that they have already lost in relation to the cultural valorisation of masculinity; femininity, even in the acceptable manifestation of a heterosexual outcome, could only represent a second-best or compromise conclusion in relation to the fantasy of completeness which turns out to have been masculine and therefore unavailable to her as a woman.

This means – as Freud says explicitly himself in his last, but far from conclusive, discussions of femininity, in the 1930s – that a girl's development to femininity is a much more arduous and less secure process than a boy's path to masculinity. Feminist readers of Freud have understood this in two ways. Either it implies that psychoanalysis explains something about why femininity should be experienced as so difficult for women to live or accept in a world that is clearly masculine through and through. Or – and this was the usual feminist view before the reconsideration which began in the mid-1970s – it makes Freud into yet one more malign male perpetrator of everything, culturally and psychically, which keeps women down and out of it.

Elizabeth Abel does not subscribe to the second of these views in any straightforward way, but she does want to hold open the possibility that what such an account may exclude is a femininity which would not be the subsidiary fall-back of a failed masculinity, but an autonomous endowment which masculine culture, with all its modes of pressure, has to work hard to exclude. Abel reads Woolf's novels for their own implied stories of female development in relation to masculine culture, looking for ways in which Woolf – without doing so knowingly – is writing a different, perhaps more womanly version of Freud's contemporary account of how sexual identities are made.

In *Virginia Woolf and the Fictions of Psychoanalysis* (1989), the book of which the piece on *Mrs Dalloway* became a part, Abel finds a marked and, for her, disappointing turn in the orientation of Woolf's narratives between the 1920s and 1930s. Abel sees the threat of fascism engendering – both in the later novels and in *Three Guineas* (1938), Woolf's polemical work on patriarchy and war – a view so dominated by the evidence around her of intractable masculine power that in effect she loses sight of the more hopeful and more feminine potential that can be found in her earlier writings. *Mrs Dalloway*, published in 1925, hints at some 'natural female values' that culture, which is masculine, and masculinity, which is not natural, forces out of the way in the course of each individual woman's story. The early female connections recovered in the novel through Clarissa Dalloway's memory of her adolescent passion for Sally Seton also suggest that the natural/cultural distinction can be attached not only to the difference between female and male, but also to that of homosexual and heterosexual: Woolf 'valorises a spontaneous homosexual love over the inhibitions of imposed heterosexuality'.

Abel maintains the hypothesis of something distinctively female, natural and homosexual which might be imagined to subvert an otherwise depressingly male-dominated story of female sexual development and general cultural values. Mary Jacobus's line is in many ways rather different, though she shares with Abel a reading method which, rather than being an application of Freud as theory to Woolf as literature, is instead interested in considering both as texts which in different ways and to different purposes take and work on possible narrative forms. Jacobus draws on Lacan and Julia Kristeva, as well as on autobiographical and other texts by Freud, to weave her way in and out of theoretical and fictional passages in their differences and crossovers. Taking issue with feminists who seek and see in the early relations of mothers and daughters a perfect harmony which is broken up only by the intrusive effects of men and their culture, she argues that this story is itself a wishful reconstruction of a kind which psychoanalysis teaches us to recognise. The state of plenitude imagined as coming before the divisions encountered by a subject apart from the mother, living in a world of signs and sexual difference, can only be thought back from this later predicament. Without the Fall having happened, in other words, the Garden of Eden could not represent the happy world of what has been lost. It would thus be mistaken for women to try to recover something which in one sense has never been; or which has come to have been only at the point of its having ceased to be. But this in no way means that the myth of wholeness should be dismissed as a simple error of fact: the implication of Jacobus's analysis is rather that it should be taken all the more seriously as just that, a myth or fantasy that is basic to human subjectivity.

Jacobus sets out in this context to show that the less-than-wholeness of subjectivity applies to both sexes, though in different ways. It makes no more sense to claim for female subjectivity a fullness missing in the male than it does to take up the notion that was once taken to be implied by Freud's concept of penis envy: that while women are a deprived sex, men unequivocally have it all. But their unsymmetrical placement within this world of meaning which is nonetheless organised around a fantasy of completeness entails that men and women will have a different relation to their sense of what they have lost or never had. For men, the operative term is threat, since they are in a position to think that they have something to lose, but for women – as in Lily Briscoe's 'to want and want and not to have', impossibly demanding the return of Mrs Ramsay – it is a kind of rootlessness which, following Kristeva and Jane Gallop, may be given the name of nostalgia.

Jacobus's Woolf novel for this excursion is *To the Lighthouse*. This has been drawn on again and again of late in discussions of mother–daughter relations and female subjectivity, as an apparently endless procession of feminist critics tiptoe their way upstairs to take one more look into that draughty bedroom where Mrs R. is forever tucking her James and Cam so gender-specifically up. There is something about this novel which seems to make it the one that feminists are always coming back to; but even in the early 1970s, Tony Inglis could note its critical predominance, and – to go back even further to the imaginary beginnings of this particular maternal story – even as early as Auerbach's chapter on Woolf, Mrs Ramsay was at the centre.

We shall come back one more time to the feminine or feminist possibilities implied by Auerbach's choices; but the next piece in the collection is one which decidedly puts *To the Lighthouse* in its less than last-word place, by making *Orlando*, the next novel chronologically, into the culmination of a process that Susan Squier refers to as Woolf's 'literary emancipation'. Rather than focusing on the pressure of mothers

– whether as havens to return to or as censorious angels in the too domestic house[12] – Squier is interested in how both *To the Lighthouse* and, in a different way, *Orlando*, involve breaking away from paternal figures. Woolf said herself that the earlier novel was her means of laying to rest the ghosts of both her parents, and she once declared that she thought that her writing had only been enabled by her father's death (she began to publish short pieces of journalism the following year, in 1905). *Orlando*, according to Squier, represents a tussle with more distant literary fathers, in particular Daniel Defoe, at once the father of the modern novel who would have to be displaced in order to make way for a distinctively feminine version of the genre and the stylistic model Woolf named when she first conceived of this novel.

In singling out *Orlando* for such honours – it is generally regarded, as she says, as the novel of light relief between the weightier consolations and experimentations of *To the Lighthouse* and *The Waves* – Squier's essay has the effect of suggesting a number of questions about the ways that both Woolf's *œuvre* and its individual texts are rated, and these in turn relate to more general questions of canon and value addressed by feminist criticism in its assessments of women writers. Squier stresses the strong femininity of *Orlando* in both inspiration and narrative development: Woolf's love for Vita Sackville-West was in some sense the book's *raison d'être*, and Orlando, in the course of her extensively unrealistic life, is advantageously altered from maleness to femaleness. At the same time, all this goes on in the playful style which has made *Orlando* appear to be so simple in its subversions. If masculine authorities in all the genres of life and literature can be undone so effortlessly and with such easy pleasure, then there might be a question about what would be the need or the use of setting up a new hierarchical order – between women writers, or between men and women writers, or between the works of an individual woman – to take the place of the old.

In quite different ways, Gillian Beer, Peggy Kamuf and Catharine Stimpson all implicitly suggest answers to this question. Beer's essay discusses how 'the advent of the aeroplane reordered the axes of experience' in relation to personal and collective conceptions of identity in the England imagined as a sealed-off island. She draws on a wide range of references, literary and non-literary, taking off from this question which at first glance might be three thousand miles away from those with which feminist criticism is concerned. And arguably, in shifting the grounds and flight-paths of debate away from such insular issues as just how good, in some narrowly literary sense, a writer is, feminist criticism – in this respect sharing its aims with those of recent Marxist and post-colonial studies – has been part of the developments which have made it possible for literary texts to be considered in all the complexity of their relations to the cultures in which they emerge.

Seen in these contexts, *The Waves*, for instance, appears not so much adrift on some timeless ocean of poetic profundity as closely anchored to the new contemporary ways of conceptualising movement implied by the science of wave motion or the development of flying machines. Innovations of narrative form – especially with regard to the self-consciousness of the perspective of an omniscient narrator who sees all from on high, but instead of treating it as natural draws attention to the exceptional oddity of such a bird's-eye, surveying view – can be related very precisely to the new kinds of seeing rendered possible by the invention of aeroplanes. The England whose proud island history is ironised in the village pageant of *Between the Acts* is thus distanced as well by the difference of view rendered inescapable by the boundary-breaking interferences of the plane which reduces the island's protective waters to redundancy. In this sense, Woolf's novels, even those that appear to be furthest from the urgencies of the present time, are shown to be inseparable from what Beer describes as the politics of *The Waves*, its 'dislimning of the boundaries of the self, the nation, the narrative'.

Woolf's interest in the relation between feminism and national politics, most fully explored in *Three Guineas* just before the war, is well known; and she was also, as Catharine Stimpson points out, interested in ways that women's historical researches and writing might shift the premises of historical undertakings. Stimpson concentrates on *A Room of One's Own* and its swinging 'glissandos about sexual difference' to show how the different views propounded at moments in that text about everything from writing to sexuality to psychology to social oppression have been followed or echoed in many of the recent variants of feminist criticism and theory.[13] This is a text which dates from the same period as *Orlando* and is every bit as full of the quality that Stimpson calls an 'ardent flair' in its writing. Stimpson does not only report this feature, but makes it part of her own style, raunchily taking up Woolf's implied invitation and exhortation to women to bring about changes as much through the ways that they do things with words as through anything which might be more recognisable as political activism.

This open, experimental practice is bound up also with a sense on Woolf's part that answers and remedies cannot be easily found; that to imagine they might be would only be to repeat the gestures of closing off – concluding and shutting out – that the narrator of *A Room* finds so difficult to tolerate in her wanderings through Oxbridge, the books of the British Library and the genre prescribed by tradition for her own lecture. As Gillian Beer puts it, 'refusing to resolve is not irresolution, but assertion'; and Stimpson carries on her version of this exploratory affirmativeness through the way that she takes the lack of settlement among the different lines of contemporary feminist criticism as a positive sign of continuing movement.

Peggy Kamuf's piece, which finishes the book, is not designed to provide that arbitrary conclusion – to Woolf criticism, to feminist theory, or even to this book. Like so many of the other pieces, one of the Woolfian notions it takes up is her inconclusiveness, as practice and as feminist policy; and here this is taken back to a kind of literal literary pre-inconclusion, with Penelope's strategic weaving and unweaving of her tapestry in order to ward off her suitors while she waits for Odysseus's return. This Homeric device is 'not a terribly clever trick', as Kamuf says, but still something which provides a woman with a kind of staying power, even as the 'weaving' of a story – a metaphor associated with the work and art of otherwise powerless women – serves as the ground for the masculine stories of war and power which exclude them in other respects. Penelope leaves us with this distinctively feminist heirloom of 'a shuttling figure in power's household, one whose movement between outside and inside, violence and poetry, the work of history and the unworking of fiction may allow us to frame one or two notions about the place of women's art'.

Penelope then becomes a kind of back-door visitor to the scenes of *A Room of One's Own*, which are equally preoccupied with relations between women, men, power, fiction and history, and the ways in which it is not simply possible to make the kind of exclusive distinctions which would enable us to say unequivocally – in conclusion – how they stand, once and for all. Bringing in the work of Michel Foucault through a different entrance, Kamuf then goes on to show how his pulling the rug from under the imagined authority of the Western humanist subject covertly preserves in the mode of its telling a certain assumption of discursive assurance – which might have something to do with Foucault's failure to engage with, or break off with, the differential ways that power operates on the two sexes. Kamuf does not seek to sew everything up with the fabrication of a new woman subject, securely installed in her different room, for that would only replicate the exclusions and false unities of the situation at the beginning. Instead, through her reading of the equivocal ins and outs of *A Room of One's Own*, she suggests that such subtle undoings of standard stories of power and subjectivity are more likely, in all their unfinished complexity, to emerge from the off-centre place of women and of femininity.

Kamuf consciously plays with the proper order of the Western literary and philosophical tradition, threading her way to and fro between Homer, Woolf and Foucault, with a brief intervention from Descartes along the way, rereading their stories with a backwards feminine glance which then retrospectively changes the sense of a history which now has to be seen as a kind of provisional fiction, always open to interruption.

Thinking along and between these lines, we might look back again at Auerbach's early case for Woolf, with its audacious promotion of the twentieth-century woman to the finishing line. For Mrs Ramsay's brown stocking – at the end of a book which begins with Odysseus's eventual return to Ithaca – now seems to figure for Woolf criticism in the same way that Penelope's weaving does for feminist criticism more generally: as a foundational feminine moment which now turns out to have changed its appearance in the light of what has happened since.

It seems in the first place ironic that Auerbach's prognosis should have found such an unexpected mode of fulfilment. Woolf does indeed now seem to many to be the central Western writer of the modern period, but for different reasons, and in the context of different concerns from those which mattered to Auerbach at the end of the Second World War. In so far as feminist critics have raised Woolf to a new position of representative glory, it is not because she comes at the end of a continuous line of development, but because she seems to have interrupted it, once and for all, perhaps to initiate new movements whose difference is marked by something provisionally identified as feminine. The literary 'emancipation' which Susan Squier applauds for Woolf herself in her writing of *Orlando* against a masculine literary tradition is figured within that novel by the hero's change of sex from man to woman, and this in a different way might seem to be emblematic of the change that has taken place in our ways of thinking about both literary and everyday transmissions of power and knowledge.

But the differences between the essays in this volume will also have made it clear that this cannot be represented as a simple progression – an unexpected dénouement or unravelling after what seemed like the end of the line, or a neat new beginning once the old fabric has been definitively worn to shreds. There is no one Woolf who can be said to have been unequivocally destined to come out in the end – for that would only mean that everything had been settled from the start, that no change had happened or would, that the conclusion was built in from the moment of birth, which by the same token would also be a moment of death. All these essays look back to draw and weave from Woolf's writing some answers and some more questions for the things that matter now, or mattered when they were written.

In a similar way, when we reread Auerbach's essay on *To the Lighthouse* we can see in it hints and possibilities which now seem to point to what has become the future of Woolf criticism, and indeed of the feminist criticism which has stitched its way now so patently into the fabric of what remains of a Western literary tradition. To the extent that he deals with the novel's authorship by a woman and its focus on female characters, Auerbach seems to be taking quite a traditional masculine view of womanliness as a mystery when he declares: 'We never come to learn what Mrs Ramsay's situation really is. Only the sadness, the vanity of her beauty and vital force emerge from the depths of secrecy.' But then he goes on to tack on a detail which changes everything: 'It is one of the few books of this type which is filled with good and genuine love but also, in its feminine way, with irony, amorphous sadness, and doubt of life.' The ironic end of Auerbach's story is feminine in its irony, then, as well as in its divergence from what had seemed to be the plan. The unfamiliar linking of sadness and irony, where both are associated with femininity, seems to foreshadow – in its own vaguely masculine way – something of the strange mixture of melancholy returns and incisive wit that mark the feminist essays in this book, in a way that seems quite appropriate to the fluctuations of Woolf's own writing. We have moved between the nostalgia of a feminine subjectivity always somewhere haunted by a lost mother and a return to the Ithaca of Penelope's ruse of unfinished weaving; between the troubles of a subjectivity that is never simply at home in one place or room, and the liberating excitement of a feminine writing which can be flying in the face of a whole tradition and trying out the multiple modes of its own possibilities. Woolf criticism is not likely to come to any final conclusions in the foreseeable future – which, as I hope to have shown, is the sign of its strength and flexibility, as the unfinished brown stocking continues to provide material for thinking with.

Notes

1. LOUISE DeSALVO, *Virginia Woolf: The Impact of Childhood Sexual Abuse on her Life and Work* (Boston: Beacon Press, 1989).

2. For more on Woolf and the essay, see my introduction to Woolf's *A Woman's Essays: Selected Essays Volume 1* (Harmondsworth: Penguin, 1992).

3. To emphasise this historical specificity, the copyright page includes the phrase: 'Written in Istanbul between May 1942 and April 1945'.

4. A detailed introduction to Woolf's reception, including reviews of her novels when they first came out, can be found in Robin Majumdar and Allen McLaurin (eds), *Virginia Woolf: The Critical Heritage* (London: Routledge & Kegan Paul, 1976).

5. *Virginia Woolf et le groupe de Bloomsbury*, ed. Jean Guiguet (Paris: Union Générale d'Editions, coll. 10/18, 1977).

6. Notably in the following exchange, in the discussion after another paper (p.138 in the French text, my translation):

   Viviane Forrester I did write to Dr Leavis, in the hope that there could be a sort of discussion between him and Quentin Bell here. He replied that he had 'nothing recordable to say about V. Woolf'. I saw this as a sign that his hostility had not softened.

Jean Guiguet Behind these attitudes, there is certainly a position that is both political and ethical. I had thought about inviting Leavis. I spoke to David Daiches about it, and he very kindly supplied me with a range of diverse positions, in a sense giving me the choice of our adversary. Prudently, I chose the least dangerous, and it is Mr Inglis, one of his [Daiches's] colleagues from Sussex University, who will represent anti-Bloomsburyism for us here.

7. See note 10, below.

8. In addition to the books of her own, Marcus edited two anthologies of feminist criticism of Woolf, *Virginia Woolf: A Feminist Slant* (1981) and *New Feminist Essays on Virginia Woolf* (1986). She has also been at the forefront of major editing projects; the special issue of the *Bulletin of the New York Public Library* in 1977, focusing on *The Years* and its early drafts edited by Mitchell Leaska as *The Pargiters*, is one example of this.

9. Virginia Woolf, 'The Leaning Tower', *A Woman's Essays*, (see note 2, above), p. 160.

10. See for instance CAROLYN G. HEILBRUN, *Towards Androgyny: Aspects of Male and Female in Literature* (1964; rpr. London: Victor Gollancz, 1973); HERBERT MARDER, *Feminism and Art: A Study of Virginia Woolf* (Chicago: University of Chicago Press, 1968); PHYLLIS ROSE, *Woman of Letters: A Life of Virginia Woolf* (London: Routledge & Kegan Paul, 1978).

11. See further DANIEL FERRER, *Virginia Woolf and the Madness of Language*, trans. Geoffrey Bennington and Rachel Bowlby (London: Routledge, 1990).

12. Woolf dramatises the figure of the Angel in the House in her essay 'Professions for Women', reprinted in *Women and Writing*, ed. Michèle Barrett (London: Women's Press, 1979), pp. 57–63, and in *The Crowded Dance of Modern Life: Selected Essays, Vol. 2*, ed. Rachel Bowlby (Harmondsworth: Penguin, 1993).

13. See further RACHEL BOWLBY, 'The Trained Mind', in *Virginia Woolf: Feminist Destinations* (Oxford: Basil Blackwell, 1988), pp. 17–48.

2 The Brown Stocking*

ERICH AUERBACH

This concluding chapter of Auerbach's classic critical work puts a novel by Woolf into a position of pre-eminence, as the culmination of nothing less than the Western literary tradition, and the sign of its possible future directions. The chapter shows what can be done through close reading of a short passage to make general remarks about the novel from which it is taken and – moving outwards – the writer's work and even a whole class of novels. What is significant in the context of subsequent developments, both for Woolf criticism and for literary theory more generally, is that not only is Auerbach's decisive novel written by a woman, but the passage in question makes a woman's daydreams the paradigm for modernist conceptions of fragment, detail and randomness. Auerbach's recommendation of Woolf – not in keeping with its time, when her reputation would not have made her seem an automatic candidate for the honour bestowed by this crucial anticipatory placement – was to find a paradoxical fulfilment in the elevation of Woolf to the highest rank, not so much as a writer in the Western tradition, but as one who broke with it as a woman.

'And even if it isn't fine to-morrow', said Mrs Ramsay, raising her eyes to glance at William Bankes and Lily Briscoe as they passed, 'it will be another day. And now', she said, thinking that Lily's charm was her Chinese eyes, aslant in her white, puckered little face, but it would take a clever man to see it, 'and now stand up, and let me measure your leg', for they might go to the Lighthouse after all, and she must see if the stocking did not need to be an inch or two longer in the leg.

Smiling, for an admirable idea had flashed upon her this very second – William and Lily should marry – she took the heather mixture

---

* Reprinted from Auerbach, *Mimesis: The Representation of Reality in Western Literature* (1946), trans. Willard R. Trask (Princeton: Princeton University Press, 1953), pp. 525–53.

stocking, with its criss-cross of steel needles at the mouth of it, and measured it against James's leg.

'My dear, stand still', she said, for in his jealousy, not liking to serve as measuring-block for the Lighthouse keeper's little boy, James fidgeted purposely; and if he did that, how could she see, was it too long, was it too short? she asked.

She looked up – what demon possessed him, her youngest, her cherished? – and saw the room, saw the chairs, thought them fearfully shabby. Their entrails, as Andrew said the other day, were all over the floor; but then what was the point, she asked herself, of buying good chairs to let them spoil up here all through the winter when the house, with only one old woman to see to it, positively dripped with wet? Never mind: the rent was precisely twopence halfpenny; the children loved it; it did her husband good to be three thousand, or if she must be accurate, three hundred miles from his library and his lectures and his disciples; and there was room for visitors. Mats, camp beds, crazy ghosts of chairs and tables whose London life of service was done – they did well enough here; and a photograph or two, and books. Books, she thought, grew of themselves. She never had time to read them. Alas! even the books that had been given her, and inscribed by the hand of the poet himself: 'For her whose wishes must be obeyed . . .' 'The happier Helen of our days . . .' disgraceful to say, she had never read them. And Croom on the Mind and Bates on the Savage Customs of Polynesia ('My dear, stand still', she said) – neither of those could one send to the Lighthouse. At a certain moment, she supposed, the house would become so shabby that something must be done. If they could be taught to wipe their feet and not bring the beach in with them – that would be something. Crabs, she had to allow, if Andrew really wished to dissect them, or if Jasper believed that one could make soup from seaweed, one could not prevent it; or Rose's objects – shells, reeds, stones; for they were gifted, her children, but all in quite different ways. And the result of it was, she sighed, taking in the whole room from floor to ceiling, as she held the stocking against James's leg, that things got shabbier and got shabbier summer after summer. The mat was fading; the wall-paper was flapping. You couldn't tell any more that those were roses on it. Still, if every door in a house is left perpetually open, and no lockmaker in the whole of Scotland can mend a bolt, things must spoil. What was the use of flinging a green Cashmere shawl over the edge of a picture frame? In two weeks it would be the colour of pea soup. But it was the doors that annoyed her; every door was left open. She listened. The drawing-room door was open; the hall door was open; it sounded as if the bedroom doors were open; and certainly the window on the landing was open, for that she had opened herself. That windows should be open, and doors shut – simple as it was, could none of them remember it? She would go into the maids' bedrooms at night and find them sealed like ovens, except for Marie's, the Swiss girl, who would rather go without a bath than without fresh air, but then at home, she had said, 'the mountains are so beautiful'. She had said that last night looking out of the window with tears in her eyes. 'The mountains are so beautiful.' Her father was dying there, Mrs Ramsay knew. He was leaving them fatherless. Scolding and demonstrating (how to make a bed, how to open a window, with hands that shut and spread like a Frenchwoman's) all had folded itself quietly about her, when the girl spoke, as, after a flight through the sunshine the wings of a bird fold themselves quietly and the blue of its plumage changes from bright steel to soft purple. She had stood there silent for there was nothing to be said. He had cancer of the throat. At the recollection – how she had stood there, how the girl had said 'At home the mountains are so beautiful', and there was no hope, no hope whatever, she had a spasm of irritation, and speaking sharply, said to James: 'Stand still. Don't be tiresome', so that he knew instantly that her severity was real, and straightened his leg and she measured it.

The stocking was too short by half an inch at least, making allowance for the fact that Sorley's little boy would be less well grown than James.

'It's too short', she said, 'ever so much too short.'

Never did anybody look so sad. Bitter and black, half-way down, in the darkness, in the shaft which ran from the sunlight to the depths, perhaps a tear formed; a tear fell; the waters swayed this way and that, received it, and were at rest. Never did anybody look so sad.

But was it nothing but looks? people said. What was there behind it – her beauty, her splendour? Had he blown his brains out, they asked, had he died the week before they were married – some other, earlier lover, of whom rumours reached one? Or was there nothing? nothing but an incomparable beauty which she lived behind, and could do nothing to disturb? For easily though she might have said at some moment of intimacy when stories of great passion, of love foiled, of ambition thwarted came her way how she too had known or felt or been through it herself, she never spoke. She was silent always. She knew then – she knew without having learnt. Her simplicity fathomed what clever people falsified. Her singleness of mind made her drop plumb like a stone, alight exact as a bird, gave her, naturally, this swoop and fall of the spirit upon truth which delighted, eased, sustained – falsely perhaps.

('Nature has but little clay', said Mr Bankes once, hearing her voice on the telephone, and much moved by it though she was only telling him a fact about a train, 'like that of which she moulded you.' He saw her at the end of the line, Greek, blue-eyed, straight-nosed. How incongruous it seemed to be telephoning to a woman like that. The Graces assembling seemed to have joined hands in meadows of asphodel to compose that face. Yes, he would catch the 10.30 at Euston.

'But she's no more aware of her beauty than a child', said Mr Bankes, replacing the receiver and crossing the room to see what progress the workmen were making with an hotel which they were building at the back of his house. And he thought of Mrs Ramsay as he looked at that stir among the unfinished walls. For always, he thought, there was something incongruous to be worked into the harmony of her face. She clapped a deerstalker's hat on her head; she ran across the lawn in galoshes to snatch a child from mischief. So that if it was her beauty merely that one thought of, one must remember the quivering thing, the living thing (they were carrying bricks up a little plank as he watched them), and work it into the picture; or if one thought of her simply as a woman, one must endow her with some freak of idiosyncrasy; or suppose some latent desire to doff her royalty of form as if her beauty bored her and all that men say of beauty, and she wanted only to be like other people, insignificant. He did not know. He did not know. He must go to his work.)

Knitting her reddish-brown hairy stocking, with her head outlined absurdly by the gilt frame, the green shawl which she had tossed over the edge of the frame, and the authenticated masterpiece by Michael Angelo, Mrs Ramsay smoothed out what had been harsh in her manner a moment before, raised his head, and kissed her little boy on the forehead. 'Let's find another picture to cut out', she said.

This piece of narrative prose is the fifth section of part 1 in Virginia Woolf's novel, *To the Lighthouse*, which was first published in 1927. The situation in which the characters find themselves can be almost completely deduced from the text itself. Nowhere in the novel is it set forth systematically, by way of introduction or exposition, or in any other way than as it is here. I shall, however, briefly summarize what the situation is at the beginning of our passage. This will make it easier for the reader to understand the following analysis; it will also serve to bring out more clearly a number of important motifs from earlier sections which are here only alluded to.

Mrs Ramsay is the wife of an eminent London professor of philosophy; she is very beautiful but definitely no longer young. With her youngest son James – he is six years old – she is sitting by the window in a good-sized summer house on one of the Hebrides islands. The professor has rented it for many years. In addition to the Ramsays, their eight children, and the servants, there are a number of guests in the house, friends on longer or shorter visits. Among them is a well-known botanist, William Bankes, an elderly widower, and Lily Briscoe, who is a painter. These two are just passing by the window. James is sitting on the floor busily cutting pictures from an illustrated catalogue. Shortly before, his mother had told him that, if the weather should be fine, they would sail to the lighthouse the next day. This is an expedition James has been looking forward to for a long time. The people at the lighthouse are to receive various presents; among these are stockings for the lighthouse-keeper's boy. The violent joy which James had felt when the trip was announced had been as violently cut short by his father's acid observation that the weather would not be fine the next day. One of the guests, with malicious emphasis, has added some corroborative meteorological details. After all the others have left the room, Mrs Ramsay, to console James, speaks the words with which our passage opens.

The continuity of the section is established through an exterior occurrence involving Mrs Ramsay and James: the measuring of the stocking. Immediately after her consoling words (if it isn't fine tomorrow, we'll go some other day), Mrs Ramsay makes James stand up so that she can measure the stocking for the lighthouse-keeper's son against his leg. A little further on she rather absent-mindedly tells him to stand still – the boy is fidgeting because his jealousy makes him a little stubborn and perhaps also because he is still under the impression of the disappointment of a few moments ago. Many lines later, the warning to stand still is repeated more sharply. James obeys, the measuring takes place, and it is found that the stocking is still considerably too short. After another long interval the scene concludes with Mrs Ramsay kissing the boy on the forehead (she thus makes up for the sharp tone of her second order to him to stand still) and her proposing to help him look for another picture to cut out. Here the section ends.

This entirely insignificant occurrence is constantly interspersed with other elements which, although they do not interrupt its progress, take up far more time in the narration than the whole scene can possibly have lasted. Most of these elements are inner processes, that is, movements within the consciousness of individual personages, and not necessarily of personages involved in the exterior occurrence but also of others who are not even present at the time: 'people', or 'Mr Bankes'. In addition other exterior occurrences which might be called secondary and which pertain to quite different times and places (the telephone conversation, the construction of the building, for example) are worked in and made to serve as the frame for what goes on in the consciousness of third persons. Let us examine this in detail.

Mrs Ramsay's very first remark is twice interrupted: first by the visual impression she receives of William Bankes and Lily Briscoe passing by together, and then, after a few intervening words serving the progress of the exterior occurrence, by the impression which the two persons passing by have left in her: the charm of Lily's Chinese eyes, which it is not for every man to see – whereupon she finishes her sentence and also allows her consciousness to dwell for a moment on the measuring of the stocking: we may yet go to the lighthouse, and so I must make sure the stocking is long enough. At this point there flashes into her mind the idea which has been prepared by her reflection on Lily's Chinese eyes (William and Lily ought to marry) – an admirable idea, she loves making matches. Smiling, she begins measuring the stocking. But the boy, in his stubborn and jealous love of her, refuses to stand still. How can she see whether the stocking is the right length if the boy keeps fidgeting about? What is the matter with James, her youngest, her darling? She looks up. Her eye falls on the room – and a long parenthesis begins. From the shabby chairs of which Andrew, her eldest son, said the other day that their entrails were all over the floor, her thoughts wander on, probing the objects and the people of her environment. The shabby furniture . . . but still good enough for up here; the advantages of the summer place; so cheap, so good for the children, for her husband; easily fitted up with a few old pieces of furniture, some pictures and books. Books – it is ages since she has had time to read books, even the books which have been dedicated to her (here the lighthouse flashes in for a second, as a place where one can't send such erudite volumes as some of those lying about the room). Then the house again: if the family would only be a little more careful. But of course, Andrew brings in crabs he wants to dissect; the other children gather seaweed, shells, stones; and she has to let them. All the children are gifted, each in a different way. But naturally, the house gets shabbier as a result (here the parenthesis is interrupted for a moment; she holds the stocking against James's leg); everything goes to ruin. If only the doors weren't always left open. See, everything is getting spoiled, even that Cashmere shawl on the picture frame. The doors are always left open; they are open again now. She listens: Yes, they are all open. The window on the landing is open too; she opened it herself. Windows must be open, doors closed. Why is it that no one can get that into his head? If you go to the maids' rooms at night, you will find all the windows closed. Only the Swiss maid always keeps her window open. She needs fresh air. Yesterday she looked out of the window with tears in her eyes and said: At home the mountains are so beautiful. Mrs Ramsay knew that 'at home' the girl's father was dying. Mrs Ramsay had just been trying to teach her how to make beds, how to open windows. She had been talking away and had scolded the girl too. But then she had stopped talking (comparison with a bird folding its wings after flying in sunlight). She had stopped talking, for there was nothing one could say; he has cancer of the throat. At this point, remembering how she had stood there, how the girl had said at home

25

the mountains were so beautiful – and there was no hope left – a sudden tense exasperation arises in her (exasperation with the cruel meaninglessness of a life whose continuance she is nevertheless striving with all her powers to abet, support, and secure). Her exasperation flows out into the exterior action. The parenthesis suddenly closes (it cannot have taken up more than a few seconds; just now she was still smiling over the thought of a marriage between Mr Bankes and Lily Briscoe), and she says sharply to James: Stand still. Don't be so tiresome.

This is the first major parenthesis. The second starts a little later, after the stocking has been measured and found to be still much too short. It starts with the paragraph which begins and ends with the motif, 'never did anybody look so sad'.

Who is speaking in this paragraph? Who is looking at Mrs Ramsay here, who concludes that never did anybody look so sad? Who is expressing these doubtful, obscure suppositions? – about the tear which – perhaps – forms and falls in the dark, about the water swaying this way and that, receiving it, and then returning to rest? There is no one near the window in the room but Mrs Ramsay and James. It cannot be either of them, nor the 'people' who begin to speak in the next paragraph. Perhaps it is the author. However, if that be so, the author certainly does not speak like one who has a knowledge of his characters – in this case, of Mrs Ramsay – and who, out of his knowledge, can describe their personality and momentary state of mind objectively and with certainty. Virginia Woolf wrote this paragraph. She did not identify it through grammatical and typographical devices as the speech or thought of a third person. One is obliged to assume that it contains direct statements of her own. But she does not seem to bear in mind that she is the author and hence ought to know how matters stand with her characters. The person speaking here, whoever it is, acts the part of one who has only an impression of Mrs Ramsay, who looks at her face and renders the impression received, but is doubtful of its proper interpretation. 'Never did anybody look so sad' is not an objective statement. In rendering the shock received by one looking at Mrs Ramsay's face, it verges upon a realm beyond reality. And in the ensuing passage the speakers no longer seem to be human beings at all but spirits between heaven and earth, nameless spirits capable of penetrating the depths of the human soul, capable too of knowing something about it, but not of attaining clarity as to what is in process there, with the result that what they report has a doubtful ring, comparable in a way to those 'certain airs, detached from the body of the wind', which in a later passage (2, 2) move about the house at night, 'questioning and wondering'. However that may be, here too we are not dealing with objective utterances on the part of the author in respect to one of the characters. No one is certain of anything here: it is all mere supposition, glances cast by one person upon another whose enigma he cannot solve.

This continues in the following paragraph. Suppositions as to the meaning of Mrs Ramsay's expression are made and discussed. But the level of tone descends slightly, from the poetic and non-real to the practical and earthly; and now a speaker is introduced: 'People said'. People wonder whether some recollection of an unhappy occurrence in her earlier life is hidden behind her radiant beauty. There have been rumors to that effect. But perhaps the rumors are wrong: nothing of this is to be learned directly from her; she is silent when such things come up in conversation. But supposing she has never experienced anything of the sort herself, she yet knows everything even without experience. The simplicity and genuineness of her being unfailingly light upon the truth of things, and, falsely perhaps, delight, ease, sustain.

Is it still 'people' who are speaking here? We might almost be tempted to doubt it, for the last words sound almost too personal and thoughtful for the gossip of 'people'. And immediately afterward, suddenly and unexpectedly, an entirely new speaker, a new scene, and a new time are introduced. We find Mr Bankes at the telephone talking to Mrs Ramsay, who has called him to tell him about a train connection, evidently with reference to a journey they are planning to make together. The paragraph about the tear had already taken us out of the room where Mrs Ramsay and James are sitting by the window; it had transported us to an undefinable scene beyond the realm of reality. The paragraph in which the rumours are discussed has a concretely earthly but not clearly identified scene. Now we find ourselves in a precisely determined place, but far away from the summer house – in London, in Mr Bankes's house. The time is not stated ('once'), but apparently the telephone conversation took place long (perhaps as much as several years) before this particular sojourn in the house on the island. But what Mr Bankes says over the telephone is in perfect continuity with the preceding paragraph. Again not objectively but in the form of the impression received by a specific person at a specific moment, it as it were sums up all that precedes – the scene with the Swiss maid, the hidden sadness in Mrs Ramsay's beautiful face, what people think about her, and the impression she makes: Nature has but little clay like that of which she molded her. Did Mr Bankes really say that to her over the telephone? Or did he only want to say it when he heard her voice, which moved him deeply, and it came into his mind how strange it was to be talking over the telephone with this wonderful woman, so like a Greek goddess? The sentence is enclosed in quotation marks, so one would suppose that he really spoke it. But this is not certain, for the first words of his soliloquy, which follows, are likewise enclosed in quotation marks. In any case, he quickly gets hold of himself, for he answers in a matter-of-fact way that he will catch the 10.30 at Euston.

But his emotion does not die away so quickly. As he puts down the receiver and walks across the room to the window in order to watch the work on a new building across the way – apparently his usual and characteristic procedure when he wants to relax and let his thoughts wander freely – he continues to be preoccupied with Mrs Ramsay. There is always something strange about her, something that does not quite go with her beauty (as for instance telephoning); she has no awareness of her beauty, or at most only a childish awareness; her dress and her actions show that at times. She is constantly getting involved in everyday realities which are hard to reconcile with the harmony of her face. In his methodical way he tries to explain her incongruities to himself. He puts forward some conjectures but cannot make up his mind. Meanwhile his momentary impressions of the work on the new building keep crowding in. Finally he gives it up. With the somewhat impatient, determined matter-of-factness of a methodical and scientific worker (which he is) he shakes off the insoluble problem 'Mrs Ramsay'. He knows no solution (the repetition of 'he did not know' symbolizes his impatient shaking it off). He has to get back to his work.

Here the second long interruption comes to an end and we are taken back to the room where Mrs Ramsay and James are. The exterior occurrence is brought to a close with the kiss on James's forehead and the resumption of the cutting out of pictures. But here too we have only an exterior change. A scene previously abandoned reappears, suddenly and with as little transition as if it had never been left, as though the long interruption were only a glance which someone (who?) has cast from it into the depths of time. But the theme (Mrs Ramsay, her beauty, the enigma of her character, her absoluteness, which nevertheless always exercises itself in the relativity and ambiguity of life, in what does not become her beauty) carries over directly from the last phase of the interruption (that is, Mr Bankes's fruitless reflections) into the situation in which we now find Mrs Ramsay: 'with her head outlined absurdly by the gilt frame', etc. – for once again what is around her is not suited to her, is 'something incongruous'. And the kiss she gives her little boy, the words she speaks to him, although they are a genuine gift of life, which James accepts as the most natural and simple truth, are yet heavy with unsolved mystery.

Our analysis of the passage yields a number of distinguishing stylistic characteristics, which we shall now attempt to formulate.

The writer as narrator of objective facts has almost completely vanished; almost everything stated appears by way of reflection in the consciousness of the dramatis personae. When it is a question of the house, for example, or of the Swiss maid, we are not given the objective information which Virginia Woolf possesses regarding these objects of her creative imagination but what Mrs Ramsay thinks or feels about them at a particular moment. Similarly we are not taken into Virginia Woolf's confidence and allowed to share her knowledge of Mrs Ramsay's character; we are given her character as it is reflected in and as it affects various figures in the novel: the nameless spirits which assume certain things about a tear, the people who wonder about her, and Mr Bankes. In our passage this goes so far that there actually seems to be no viewpoint at all outside the novel from which the people and events within it are observed, any more than there seems to be an objective reality apart from what is in the consciousness of the characters. Remnants of such a reality survive at best in brief references to the exterior frame of the action, such as 'said Mrs Ramsay, raising her eyes . . .' or 'said Mr Bankes once, hearing her voice'. The last paragraph ('Knitting her reddish-brown hairy stocking . . .') might perhaps also be mentioned in this connection. But this is already somewhat doubtful. The occurrence is described objectively, but as for its interpretation, the tone indicates that the author looks at Mrs Ramsay not with knowing but with doubting and questioning eyes – even as some character in the novel would see her in the situation in which she is described, would hear her speak the words given.

The devices employed in this instance (and by a number of contemporary writers as well) to express the contents of the consciousness of the dramatis personae have been analyzed and described syntactically. Some of them have been named (*erlebte Rede*, stream of consciousness, *monologue intérieur* are examples). Yet these stylistic forms, especially the *erlebte Rede*, were used in literature much earlier too, but not for the same aesthetic purpose. And in addition to them there are other possibilities – hardly definable in terms of syntax – of obscuring and even obliterating the impression of an objective reality completely known to the author; possibilities, that is, dependent not on form but on intonation and context. A case in point is the passage under discussion, where the author at times achieves the intended effect by representing herself to be someone who doubts, wonders, hesitates, as though the truth about her characters were not better known to her than it is to them or to the reader. It is all, then, a matter of the author's attitude toward the reality of the world he represents. And this attitude differs entirely from that of authors who interpret the actions, situations, and characters of their personages with objective assurance, as was the general practice in earlier times. Goethe or Keller, Dickens or Meredith, Balzac or Zola told us out of their certain knowledge what their characters did, what they felt and thought while doing it, and how their actions and thoughts were to be interpreted. They knew everything about their characters. To be sure, in past periods too we were frequently told about the subjective reactions of the characters in a novel or story; at times even in the form of *erlebte Rede*, although more frequently as a monologue, and of course in most instances with an introductory phrase something like 'it seemed to him that . . .' or 'at this moment he felt that . . .' or the like. Yet in such cases there was hardly ever any attempt to render the flow and the play of consciousness adrift in the current of changing impressions (as is done in our text both for Mrs Ramsay and for Mr Bankes); instead, the content of the individual's consciousness was rationally limited to things connected with the particular incident being related or the particular situation being described. And what is still more important: the author, with his knowledge of an objective truth, never abdicated his position as the final and governing authority. Again, earlier writers, especially from the end of the nineteenth century on, had produced narrative works which on the whole undertook to give us an extremely subjective, individualistic, and often eccentrically aberrant impression of reality, and which neither sought nor were able to ascertain anything objective or generally valid in regard to it. Sometimes such works took the form of first-person novels; sometimes they did not. As an example of the latter case I mention Huysmans's novel *A rebours*. But all that too is basically different from the modern procedure here described on the basis of Virginia Woolf's text, although the latter, it is true, evolved from the former. The essential characteristic of the technique represented by Virginia Woolf is that we are given not merely one person whose consciousness (that is, the impressions it receives) is rendered, but many persons, with frequent shifts from one to the other – in our text, Mrs Ramsay, 'people', Mr Bankes, in brief interludes James, the Swiss maid in a flash-back, and the nameless ones who speculate over a tear. The multiplicity of persons suggests that we are here after all confronted with an endeavor to investigate an objective reality, that is, specifically, the 'real' Mrs Ramsay. She is, to be sure, an enigma and such she basically remains, but she is as it were encircled by the content of all the various consciousnesses directed upon her (including her own); there is an attempt to approach her from many sides as closely as human possibilities of perception and expression can succeed in doing. The design of a close approach to objective reality by means of numerous subjective impressions received by various individuals (and at various times) is important in the modern technique which we are here examining. It basically differentiates it from the unipersonal subjectivism which allows only a single and generally a very unusual person to make himself heard and admits only that one person's way of looking at reality. In terms of literary history, to be sure, there are close connections between the two methods of representing consciousness – the unipersonal subjective method and the multipersonal method with synthesis as its aim. The latter developed from the former, and there are works in which the two overlap, so that we can watch the development. This is especially the case in Marcel Proust's great novel. We shall return to it later.

Another stylistic peculiarity to be observed in our text – though one that is closely and necessarily connected with the 'multipersonal representation of consciousness' just discussed – has to do with the treatment of time. That there is something peculiar about the treatment of time in modern narrative literature is nothing new; several studies have been published on the subject. These were primarily attempts to establish a connection between the pertinent phenomena and contemporary philosophical doctrines or trends – undoubtedly a justifiable undertaking and useful for an appreciation of the community of interests and inner purposes shown in the activity of many of our contemporaries. We shall begin by describing the procedure with reference to our present example. We remarked earlier that the act of measuring the length of the stocking and the speaking of the words related to it must have taken much less time than an attentive reader who tries not to miss anything will require to read the passage – even if we assume that a brief pause intervened between the measuring and the kiss of reconciliation on James's forehead. However, the time the narration takes is not devoted to the occurrence itself (which is rendered rather tersely), but to interludes. Two long excursuses are inserted, whose relations in time to the occurrence which frames them seem to be entirely different. The first excursus, a representation of what goes on in Mrs Ramsay's mind while she measures the stocking (more precisely, between the first absent-minded and the second sharp order to James to hold his leg still) belongs in time to the framing occurrence, and it is only the representation of it which takes a greater number of seconds and even minutes than the measuring – the reason being that the road taken by consciousness is sometimes traversed far more quickly than language is able to render it, if we want to make ourselves intelligible to a third person, and that is the intention here. What goes on in Mrs Ramsay's mind in itself contains nothing enigmatic; these are ideas which arise from her daily life and may well be called normal – her secret lies deeper, and it is only when the switch from the open windows to the Swiss maid's words comes, that something happens which lifts the veil a little. On the whole, however, the mirroring of Mrs Ramsay's consciousness is much more easily comprehensible than the sort of thing we get in such cases from other authors (James Joyce, for example). But simple and trivial as are the ideas which arise one after the other in Mrs Ramsay's consciousness, they are at the same time essential and significant. They amount to a synthesis of the intricacies of life in which her incomparable beauty has been caught, in which it at once manifests and conceals itself. Of course, writers of earlier periods too occasionally devoted some time and a few sentences to telling the reader what at a specific moment passed through their characters' minds – but for such a purpose they would hardly have chosen so accidental an occasion as Mrs Ramsay's looking up, so that, quite involuntarily, her eyes fall on the furniture. Nor would it have occurred to them to render the continuous rumination of consciousness in its natural and purposeless freedom. And finally they would not have inserted the entire process between two exterior occurrences so close together in time as the two warnings to James to keep still (both of which, after all, take place while she is on the point of holding the unfinished stocking to his leg); so that, in a surprising fashion unknown to earlier periods, a sharp contrast results between the brief span of time occupied by the exterior event and the dreamlike wealth of a process of consciousness which traverses a whole subjective universe. These are the characteristic and distinctively new features of the technique: a chance occasion releasing processes of consciousness; a natural and even, if you will, a naturalistic rendering of those processes in their peculiar freedom, which is neither restrained by a purpose nor directed by a specific subject of thought; elaboration of the contrast between 'exterior' and 'interior' time. The three have in common what they reveal of the author's attitude: he submits, much more than was done in earlier realistic works, to the random contingency of real phenomena; and even though he winnows and stylises the material of the real world – as of course he cannot help doing – he does not proceed rationalistically, nor with a view to bringing a continuity of exterior events to a planned conclusion. In Virginia Woolf's case the exterior events have actually lost their hegemony, they serve to release and interpret inner events, whereas before her time (and still today in many instances) inner movements preponderantly function to prepare and motivate significant exterior happenings. This too is apparent in the randomness and contingency of the exterior occasion (looking up because James does not keep his foot still), which releases the much more significant inner process.

The temporal relation between the second excursus and the framing occurrence is of a different sort: its content (the passage on the tear, the things people think about Mrs Ramsay, the telephone conversation with Mr Bankes and his reflections while watching the building of the new hotel) is not a part of the framing occurrence either in terms of time or of place. Other times and places are in question; it is an excursus of the same type as the story of the origin of Odysseus' scar, which was discussed in the first chapter of this book. Even from that, however, it is different in structure. In the Homer passage the excursus was linked to the scar which Euryclea touches with her hands, and although the moment at which the touching of the scar occurs is one of high and dramatic tension, the scene nevertheless immediately shifts to another clear and luminous present, and this present seems actually designed to cut off the dramatic tension and cause the entire footwashing scene to be temporarily forgotten. In Virginia Woolf's passage, there is no question of any tension. Nothing of importance in a dramatic sense takes place; the problem is the length of the stocking. The point of departure for the excursus is Mrs Ramsay's facial expression: 'never did anybody look so sad'. In fact several excursuses start from here; three, to be exact. And all three differ in time and place, differ too in definiteness of time and place, the first being situated quite vaguely, the second somewhat more definitely, and the third with comparative precision. Yet none of them is so exactly situated in time as the successive episodes of the story of Odysseus' youth, for even in the case of the telephone scene we have only an inexact indication of when it occurred. As a result it becomes possible to accomplish the shifting of the scene away from the windownook much more unnoticeably and smoothly than the changing of scene and time in the episode of the scar. In the passage on the tear the reader may still be in doubt as to whether there has been any shift at all. The nameless speakers may have entered the room and be looking at Mrs Ramsay. In the second paragraph this interpretation is no longer possible, but the 'people' whose gossip is reproduced are still looking at Mrs Ramsay's face – not here and now, at the summer-house window, but it is still the same face and has the same expression. And even in the third part, where the face is no longer physically seen (for Mr Bankes is talking to Mrs Ramsay over the telephone), it is nonetheless present to his inner vision; so that not for an instant does the theme (the solution of the enigma Mrs Ramsay), and even the moment when the problem is formulated (the expression of her face while she measures the length of the stocking), vanish from the reader's memory. In terms of the exterior event the three parts of the excursus have nothing to do with one another. They have no common and externally coherent development, as have the episodes of Odysseus' youth which are related with reference to the origin of the scar; they are connected only by the one thing they have in common – looking at Mrs Ramsay, and more specifically at the Mrs Ramsay who, with an unfathomable expression of sadness behind her radiant beauty, concludes that the stocking is still much too short. It is only this common focus which connects the otherwise totally different parts of the excursus; but the connection is strong enough to deprive them of the independent 'present' which the episode of the scar possesses. They are nothing but attempts to interpret 'never did anybody look so sad'; they carry on this theme, which itself carries on after they conclude: there has been no change of theme at all. In contrast, the scene in which Euryclea recognizes Odysseus is interrupted and divided into two parts by the excursus on the origin of the scar. In our passage, there is no such clear distinction between two exterior occurrences and between two presents. However insignificant as an exterior event the framing occurrence (the measuring of the stocking) may be, the picture of Mrs Ramsay's face which arises from it remains present throughout the excursus; the excursus itself is nothing but a background for that picture, which seems as it were to open into the depths of time – just as the first excursus, released by Mrs Ramsay's unintentional glance at the furniture, was an opening of the picture into the depths of consciousness.

The two excursuses, then, are not as different as they at first appeared. It is not so very important that the first, so far as time is concerned (and place too), runs its course within the framing occurrence, while the second conjures up other times and places. The times and places of the second are not independent; they serve only the polyphonic treatment of the image which releases it; as a matter of fact, they impress us (as does the interior time of the first excursus) like an occurrence in the consciousness of some observer (to be sure, he is not identified) who might see Mrs Ramsay at the described moment and whose meditation upon the unsolved enigma of her personality might contain memories of what others (people, Mr Bankes) say and think about her. In both excursuses we are dealing with attempts to fathom a more genuine, a deeper, and indeed a more real reality; in both cases the incident which releases the excursus appears accidental and is poor in content; in both cases it makes little difference whether the excursuses employ only the consciousness-content, and hence only interior time, or whether they also employ exterior shifts of time. After all, the process of consciousness in the first excursus likewise includes shifts of time and scene, especially the episode with the Swiss maid. The important point is that an insignificant exterior occurrence releases ideas and chains of ideas which cut loose from the present of the exterior occurrence and range freely through the depths of time. It is as though an apparently simple text revealed its proper content only in the commentary on it, a simple musical theme only in the development-section. This enables us also to understand the close relation between the treatment of time and the 'multi-personal representation of consciousness' discussed earlier. The ideas arising in consciousness are not tied to the present of the exterior occurrence which releases them. Virginia Woolf's peculiar technique, as exemplified in our text, consists in the fact that the exterior objective reality of the momentary present which the author directly reports and which appears as established fact – in our instance the measuring of the stocking – is nothing but an occasion (although perhaps not an entirely accidental one). The stress is placed entirely on what the occasion releases, things which are not seen directly but by reflection, which are not tied to the present of the framing occurrence which releases them.

Here it is only natural that we should recall Proust's work. He was the first to carry this sort of thing through consistently; and his entire technique is bound up with a recovery of lost realities in remembrance, a recovery released by some externally insignificant and apparently accidental occurrence. Proust describes the procedure more than once. We have to wait until volume 2 of *Le Temps retrouvé* for a full description embracing the corresponding theory of art; but the first description, which occurs as early as section 1 of *Du Côté de chez Swann*, is impressive enough. Here, one unpleasant winter evening, the taste of a cake (*petite Madeleine*) dipped in tea arouses in the narrator an overwhelming though at first indefinite delight. By intense and repeated effort he attempts to fathom its nature and cause, and it develops that the delight is based on a recovery: the recovery of the taste of the *petite Madeleine* dipped in tea which his aunt would give him on Sundays when, still a little boy, he went into her room to wish her good morning, in the house in the old provincial town of Combray where she lived, hardly ever leaving her bed, and where he used to spend the summer months with his parents. And from this recovered remembrance, the world of his childhood emerges into light, becomes depictable, as more genuine and more real than any experienced present – and he begins to narrate. Now with Proust a narrating 'I' is preserved throughout. It is not, to be sure, an author observing from without but a person involved in the action and pervading it with the distinctive flavor of his being, so that one might feel tempted to class Proust's novel among the products of the unipersonal subjectivism which we discussed earlier. So to class it would not be wrong but it would be inadequate. It would fail to account completely for the structure of Proust's novel. After all, it does not display the same strictly unipersonal approach to reality as Huysmans's *A rebours* or Knut Hamsun's *Pan* (to mention two basically different examples which are yet comparable in this respect). Proust aims at objectivity, he wants to bring out the essence of events: he strives to attain this goal by accepting the guidance of his own consciousness – not, however, of his consciousness as it happens to be at any particular moment but as it remembers things. A consciousness in which remembrance causes past realities to arise, which has long since left behind the states in which it found itself when those realities occurred as a present, sees and arranges that content in a way very different from the purely individual and subjective. Freed from its various earlier involvements, consciousness views its own past layers and their content in perspective; it keeps confronting them with one another, emancipating them from their exterior temporal continuity as well as from the narrow meanings they seemed to have when they were bound to a particular present. There is to be noted in this a fusion of the modern concept of interior time with the neo-Platonic idea that the true prototype of a given subject is to be found in the soul of the artist; in this case, of an artist who, present in the subject itself, has detached himself from it as observer and thus comes face to face with his own past.

I shall here give a brief passage from Proust in order to illustrate this point. It deals with a moment in the narrator's childhood and occurs in volume 1, toward the end of the first section. It is, I must admit, too good and too clear an example of the layered structure of a consciousness engaged in recollection. That structure is not always as evident as it is in this instance; elsewhere it could be made clearly apparent only through an analysis of the way the subject matter is arranged, of the introduction, disappearance, and reappearance of the characters, and of the overlapping of the various presents and consciousness-contents. But every reader of Proust will admit that the whole work is written in accordance with the technique which our passage makes apparent without comment or analysis. The situation is this: One evening during his childhood the narrator could not go to sleep without the usual ceremony of being kissed good night by his mother. When he went to bed his mother could not come to his room because there was a guest for supper. In a state of nervous hypertension he decides to stay awake and catch his mother at the door when, after the guest's departure, she herself retires. This is a serious offense, because his parents are trying to correct his excessive sensitivity by sternly suppressing such cravings. He has to reckon with severe punishment; perhaps he will be banished from home and sent to a boarding school. Yet his need for momentary satisfaction is stronger than his fear of the consequences. Quite unexpectedly it happens that his father, who is usually far stricter and more authoritarian but at the same time less consistent than his mother, comes upstairs directly behind her. Seeing the boy, he is touched by the desperate expression in his face and advises his wife to spend the night in the child's room to calm him down. Proust continues:

On ne pouvait pas remercier mon père; on l'eût agacé par ce qu'il appelait des sensibleries. Je restai sans oser faire un mouvement; il était encore devant nous, grand, dans sa robe de nuit blanche sous le cachemire de l'Inde violet et rose qu'il nouait autour de sa tête depuis qu'il avait des névralgies, avec le geste d'Abraham dans la gravure d'après Benozzo Gozzoli que m'avait donné M. Swann, disant à Hagar, qu'elle a à se départir du côté d'Isaac. Il y a bien des années de cela. La muraille de l'escalier, où je vis monter le reflet de sa bougie n'existe plus depuis longtemps. En moi aussi bien des choses ont été détruites que je croyais devoir durer toujours et de nouvelles se sont édifiées donnant naissance à des peines et à des joies nouvelles que je n'aurais pu prévoir alors, de même que les anciennes me sont devenues difficiles à comprendre. Il y a bien longtemps aussi que mon père a cessé de pouvoir dire à maman: 'Va avec le petit.' La possibilité de telles heures ne renaîtra jamais pour moi. Mais depuis peu de temps, je recommence à très bien percevoir si je prête l'oreille, les sanglots que j'eus la force de contenir devant mon père et qui n'éclatèrent que quand je me retrouvai seul avec maman. En réalité ils n'ont jamais cessé; et c'est seulement parce que la vie se tait maintenant davantage autour de moi que je les entends de nouveau, comme ces cloches de couvents que couvrent si bien les bruits de la ville pendant le jour qu'on les croirait arrêtées mais qui se remettent à sonner dans le silence du soir.

It was impossible for me to thank my father; what he called my sentimentality would have exasperated him. I stood there, not daring to move; he was still confronting us, an immense figure in his white nightshirt, crowned with the pink and violet scarf of Indian cashmere in which, since he had begun to suffer from neuralgia, he used to tie up his head, standing like Abraham in the engraving after Benozzo Gozzoli which M. Swann had given me, telling Hagar that she must tear herself away from Isaac. Many years have passed since that night. The wall of the staircase, up which I had watched the light of his candle gradually climb, was long ago demolished. And in myself, too, many things have perished which, I imagined, would last for ever, and new structures have arisen, giving birth to new sorrows and new joys which in those days I could not have foreseen, just as now the old are difficult of comprehension. It is a long time, too, since my father has been able to tell Mamma to 'Go with the child.' Never again will such hours be possible for me. But of late I have been increasingly able to catch, if I listen attentively, the sound of the sobs which I had the strength to control in my father's presence, and which broke out only when I found myself alone with Mamma. Actually, their echo has never ceased: it is only because life is now growing more and more quiet round about me that I hear them afresh, like those convent bells which are so effectively drowned during the day by the noises of the streets that one would suppose them to have been stopped for ever, until they sound out again through the silent evening air.

(Marcel Proust, *Remembrance of Things Past*, trans. C. K. Scott Moncrieff, Random House, 1934)

Through the temporal perspective we sense here an element of the symbolic omnitemporality of an event fixed in a remembering consciousness. Still clearer and more systematic (and also, to be sure, much more enigmatic) are the symbolic references in James Joyce's *Ulysses*, in which the technique of a multiple reflection of consciousness and of multiple time strata would seem to be employed more radically than anywhere else. The book unmistakably aims at a symbolic synthesis of the theme 'Everyman'. All the great motifs of the cultural history of Europe are contained in it, although its point of departure is very specific individuals and a clearly established present (Dublin, 16 June, 1904). On sensitive readers it can produce a very strong immediate impression. Really to understand it, however, is not an easy matter, for it makes severe demands on the reader's patience and learning by its dizzying whirl of motifs, wealth of words and concepts, perpetual playing upon their countless associations, and the ever rearoused but never satisfied doubt as to what order is ultimately hidden behind so much apparent arbitrariness.

Few writers have made so consistent a use of reflected consciousness and time strata as those we have so far discussed. But the influence of the procedure and traces of it can be found almost everywhere – lately even in writers of the sort whom discriminating readers are not in the habit of regarding as fully competent. Many writers have invented their own methods – or at least have experimented in the direction – of making the reality which they adopt as their subject appear in changing lights and changing strata, or of abandoning the specific angle of observation of either a seemingly objective or purely subjective representation in favor of a more varied perspective. Among these writers we find older masters whose aesthetic individualities had long since been fully established, but who were drawn into the movement in their years of maturity before and after the First World War, each in his own way turning to a disintegration and dissolution of external realities for a richer and more essential interpretation of them. Thomas Mann is an example, who, ever since his *Magic Mountain*, without in any way abandoning his level of tone (in which the narrating, commenting, objectivizing author addressing the reader is always present) has been more and more concerned with time perspectives and the symbolic omnitemporality of events. Another very different instance is André Gide, in whose *Faux-Monnayeurs* there is a constant shifting of the viewpoint from which the events (themselves multilayered) are surveyed, and who carries this procedure to such an extreme that the novel and the account of the genesis of the novel are interwoven in the ironic vein of the romanticists. Very different again, and much simpler, is the case of Knut Hamsun who, for example in his *Growth of the Soil*, employs a level of tone which blurs the dividing line between the direct or indirect discourse of the characters in the novel and the author's own utterances; as a result one is never quite certain that what one hears is being said by the author as he stands outside his novel; the statements sound as though they came from one of the persons involved in the action, or at least from a passer-by who observes the incident. Finally, we have still to mention certain further peculiarities of the kind of writing we are considering – those which concern the type of subject matter treated.

In modern novels we frequently observe that it is not one person or a limited number of persons whose experiences are pursued as a continuum; indeed, often there is no strict continuum of events. Sometimes many individuals, or many fragments of events, are loosely joined so that the reader has no definite thread of action which he can always follow. There are novels which attempt to reconstruct a milieu from mere splinters of events, with constantly changing though occasionally reappearing characters. In this latter case one might feel inclined to assume that it was the writer's purpose to exploit the structural possibilities of the film in the interest of the novel. If so, it is a wrong direction: a concentration of space and time such as can be achieved by the film (for example the representation, within a few seconds and by means of a few pictures, of the situation of a widely dispersed group of people, of a great city, an army, a war, an entire country) can never be within the reach of the spoken or written word. To be sure, the novel possesses great freedom in its command of space and time – much more than the drama of pre-film days, even if we disregard the strict classical rules of unity. The novel in recent decades has made use of this freedom in a way for which earlier literary periods afford no models, with the possible exception of a few tentative efforts by the romanticists, especially in Germany, although they did not restrict themselves to the material of reality. At the same time, however, by virtue of the film's existence, the novel has come to be more clearly aware than ever before of the limitations in space and time imposed upon it by its instrument: language. As a result the situation has been reversed: the dramatic technique of the film now has far greater possibilities in the direction of condensing time and space than has the novel itself.

The distinctive characteristics of the realistic novel of the era between the two great wars, as they have appeared in the present chapter – multipersonal representation of consciousness, time strata, disintegration of the continuity of exterior events, shifting of the narrative viewpoint (all of which are interrelated and difficult to separate) – seem to us indicative of a striving for certain objectives, of certain tendencies and needs on the part of both authors and public. These objectives, tendencies, and needs are numerous; they seem in part to be mutually contradictory; yet they form so much one whole that when we undertake to describe them analytically, we are in constant danger of unwittingly passing from one to another.

Let us begin with a tendency which is particularly striking in our text from Virginia Woolf. She holds to minor, unimpressive, random events: measuring the stocking, a fragment of a conversation with the maid, a telephone call. Great changes, exterior turning points, let alone catastrophes, do not occur; and though elsewhere in *To the Lighthouse* such things are mentioned, it is hastily, without preparation or context, incidentally, and as it were only for the sake of information. The same tendency is to be observed in other and very different writers, such as Proust or Hamsun. In Thomas Mann's *Buddenbrooks* we still have a novel structure consisting of the chronological sequence of important exterior events which affect the Buddenbrook family; and if Flaubert – in many respects a precursor – lingers as a matter of principle over insignificant events and everyday circumstances which hardly advance the action, there is nevertheless to be sensed throughout *Madame Bovary* (though we may wonder how this would have worked out in *Bouvard et Pécuchet*) a constant slow-moving chronological approach first to partial crises and finally to the concluding catastrophe, and it is this approach which dominates the plan of the work as a whole. But a shift in emphasis followed; and now many writers present minor happenings, which are insignificant as exterior factors in a person's destiny, for their own sake or rather as points of departure for the development of motifs, for a penetration which opens up new perspectives into a milieu or a consciousness or the given historical setting. They have discarded presenting the story of their characters with any claim to exterior completeness, in chronological order, and with the emphasis on important exterior turning points of destiny. James Joyce's tremendous novel – an encyclopedic work, a mirror of Dublin, of Ireland, a mirror too of Europe and its millennia – has for its frame the externally insignificant course of a day in the lives of a schoolteacher and an advertising broker. It takes up less than twenty-four hours in their lives – just as *To the Lighthouse* describes portions of two days widely separated in time. (There is here also, as we must not fail to observe, a similarity to Dante's *Comedy*.) Proust presents individual days and hours from different periods, but the exterior events which are the determining factors in the destinies of the novel's characters during the intervening lapses of time are mentioned only incidentally, in retrospect or anticipation. The ends the narrator has in mind are not to be seen in them; often the reader has to supplement them. The way in which the father's death is brought up in the passage cited above – incidentally, allusively, and in anticipation – offers a good example. This shift of emphasis expresses something that we might call a transfer of confidence: the great exterior turning points and blows of fate are granted less importance; they are credited with less power of yielding decisive information concerning the subject; on the other hand there is confidence that in any random fragment plucked from the course of a life at any time the totality of its fate is contained and can be portrayed. There is greater confidence in syntheses gained through full exploitation of an everyday occurrence than in a chronologically well-ordered total treatment which accompanies the subject from beginning to end, attempts not to omit anything externally important, and emphasizes the great turning points of destiny. It is possible to compare this technique of modern writers with that of certain modern philologists who hold that the interpretation of a few passages from *Hamlet*, *Phèdre*, or *Faust* can be made to yield more, and more decisive, information about Shakespeare, Racine, or Goethe and their times than would a systematic and chronological treatment of their lives and works. Indeed, the present book may be cited as an illustration. I could never have written anything in the nature of a history of European realism; the material would have swamped me; I should have had to enter into hopeless discussions concerning the delimitation of the various periods and the allocation of the various writers to them, and above all concerning the definition of the concept realism. Furthermore, for the sake of completeness, I should have had to deal with some things of which I am but casually informed, and hence to become acquainted with them *ad hoc* by reading up on them (which, in my opinion, is a poor way of acquiring and using knowledge); and the motifs which direct my investigation, and for the sake of which it is written, would have been completely buried under a mass of factual information which has long been known and can easily be looked up in reference books. As opposed to this I see the possibility of success and profit in a method which consists in letting myself be guided by a few motifs which I have worked out gradually and without a specific purpose, and in trying them out on a series of texts which have become familiar and vital to me in the course of my philological activity; for I am convinced that these basic motifs in the history of the representation of reality – provided I have seen them correctly – must be demonstrable in any random realistic text. But to return to those modern writers who prefer the exploitation of random everyday events, contained within a few hours and days, to the complete and chronological representation of a total exterior continuum – they too (more or less consciously) are guided by the consideration that it is a hopeless venture to try to be really complete within the total exterior continuum and yet to make what is essential stand out. Then too they hesitate to impose upon life, which is their subject, an order which it does not possess in itself. He who represents the course of a human life, or a sequence of events extending over a prolonged period of time, and represents it from beginning to end, must prune and isolate arbitrarily. Life has always long since begun, and it is always still going on. And the people whose story the author is telling experience much more than he can ever hope to tell. But the things that happen to a few individuals in the course of a few minutes, hours, or possibly even days – these one can hope to report with reasonable completeness. And here, furthermore, one comes upon the order and the interpretation of life which arise from life itself: that is, those which grow up in the individuals themselves, which are to be discerned in their thoughts, their consciousness, and in a more concealed form in their words and actions. For there is always going on within us a process of formulation and interpretation whose subject matter is our own self. We are constantly endeavoring to give meaning and order to our lives in the past, the present, and the future, to our surroundings, the world in which we live; with the result that our lives appear in our own conception as total entities – which to be sure are always changing, more or less radically, more or less rapidly, depending on the extent to which we are obliged, inclined, and able to assimilate the onrush of new experience. These are the forms of order and interpretation which the modern writers here under discussion attempt to grasp in the random moment – not one order and one interpretation, but many, which may either be those of different persons or of the same person at different times; so that overlapping, complementing, and contradiction yield something that we might call a synthesized cosmic view or at least a challenge to the reader's will to interpretive synthesis.

Here we have returned once again to the reflection of multiple consciousnesses. It is easy to understand that such a technique had to develop gradually and that it did so precisely during the decades of the First World War period and after. The widening of man's horizon, and the increase of his experiences, knowledge, ideas, and possible forms of existence, which began in the sixteenth century, continued through the nineteenth at an ever faster tempo – with such a tremendous acceleration since the beginning of the twentieth that synthetic and objective attempts at interpretation are produced and demolished every instant. The tremendous tempo of the changes proved the more confusing because they could not be surveyed as a whole. They occurred simultaneously in many separate departments of science, technology, and economics, with the result that no one – not even those who were leaders in the separate departments – could foresee or evaluate the resulting overall situations. Furthermore, the changes did not produce the same effects in all places, so that the differences of attainment between the various social strata of one and the same people and between different peoples came to be – if not greater – at least more noticeable. The spread of publicity and the crowding of mankind on a shrinking globe sharpened awareness of the differences in ways of life and attitudes, and mobilized the interests and forms of existence which the new changes either furthered or threatened. In all parts of the world crises of adjustment arose; they increased in number and coalesced. They led to the upheavals which we have not weathered yet. In Europe this violent clash of the most heterogeneous ways of life and kinds of endeavor undermined not only those religious, philosophical, ethical and economic principles which were part of the traditional heritage and which, despite many earlier shocks, had maintained their position of authority through slow adaptation and transformation; nor yet only the ideas of the Enlightenment, the ideas of

democracy and liberalism which had been revolutionary in the
eighteenth century and were still so during the first half of the
nineteenth; it undermined even the new revolutionary forces of
socialism, whose origins did not go back beyond the heyday of the
capitalist system. These forces threatened to split up and disintegrate.
They lost their unity and clear definition through the formation of
numerous mutually hostile groups, through strange alliances which
some of these groups effected with non-socialist ideologies, through the
capitulation of most of them during the First World War, and finally
through the propensity on the part of many of their most radical
advocates for changing over into the camp of their most extreme
enemies. Otherwise too there was an increasingly strong factionalism – at
times crystallizing around important poets, philosophers and scholars,
but in the majority of cases pseudo-scientific, syncretistic and primitive.
The temptation to entrust oneself to a sect which solved all problems
with a single formula, whose power of suggestion imposed solidarity,
and which ostracized everything which would not fit in and submit – this
temptation was so great that, with many people, fascism hardly had to
employ force when the time came for it to spread through the countries
of old European culture, absorbing the smaller sects.

As recently as the nineteenth century, and even at the beginning of the
twentieth, so much clearly formulable and recognized community of
thought and feeling remained in those countries that a writer engaged in
representing reality had reliable criteria at hand by which to organize it.
At least, within the range of contemporary movements, he could discern
certain specific trends; he could delimit opposing attitudes and ways of
life with a certain degree of clarity. To be sure, this had long since begun
to grow increasingly difficult. Flaubert (to confine ourselves to realistic
writers) already suffered from the lack of valid foundations for his work;
and the subsequent increasing predilection for ruthlessly subjectivistic
perspectives is another symptom. At the time of the First World War and
after – in a Europe unsure of itself, overflowing with unsettled ideologies
and ways of life, and pregnant with disaster – certain writers
distinguished by instinct and insight find a method which dissolves
reality into multiple and multivalent reflections of consciousness. That
this method should have been developed at this time is not hard to
understand.

But the method is not only a symptom of the confusion and
helplessness, not only a mirror of the decline of our world. There is, to
be sure, a good deal to be said for such a view. There is in all these works
a certain atmosphere of universal doom: especially in *Ulysses*, with its
mocking *odi-et-amo* hodgepodge of the European tradition, with its
blatant and painful cynicism, and its uninterpretable symbolism – for
even the most painstaking analysis can hardly emerge with anything

43

more than an appreciation of the multiple enmeshment of the motifs but with nothing of the purpose and meaning of the work itself. And most of the other novels which employ multiple reflection of consciousness also leave the reader with an impression of hopelessness. There is often something confusing, something hazy about them, something hostile to the reality which they represent. We not infrequently find a turning away from the practical will to live, or delight in portraying it under its most brutal forms. There is hatred of culture and civilization, brought out by means of the subtlest stylistic devices which culture and civilization have developed, and often a radical and fanatical urge to destroy. Common to almost all of these novels is haziness, vague indefinability of meaning: precisely the kind of uninterpretable symbolism which is also to be encountered in other forms of art of the same period.

But something entirely different takes place here too. Let us turn again to the text which was our starting-point. It breathes an air of vague and hopeless sadness. We never come to learn what Mrs Ramsay's situation really is. Only the sadness, the vanity of her beauty and vital force emerge from the depths of secrecy. Even when we have read the whole novel, the meaning of the relationship between the planned trip to the lighthouse and the actual trip many years later remains unexpressed, enigmatic, only dimly to be conjectured, as does the content of Lily Briscoe's concluding vision which enables her to finish her painting with one stroke of the brush. It is one of the few books of this type which are filled with good and genuine love but also, in its feminine way, with irony, amorphous sadness, and doubt of life. Yet what realistic depth is achieved in every individual occurrence, for example the measuring of the stocking! Aspects of the occurrence come to the fore, and links to other occurrences, which, before this time, had hardly been sensed, which had never been clearly seen and attended to, and yet they are determining factors in our real lives. What takes place here in Virginia Woolf's novel is precisely what was attempted everywhere in works of this kind (although not everywhere with the same insight and mastery) – that is, to put the emphasis on the random occurrence, to exploit it not in the service of a planned continuity of action but in itself. And in the process something new and elemental appeared: nothing less than the wealth of reality and depth of life in every moment to which we surrender ourselves without prejudice. To be sure, what happens in that moment – be it outer or inner processes – concerns in a very personal way the individuals who live in it, but it also (and for that very reason) concerns the elementary things which men in general have in common. It is precisely the random moment which is comparatively independent of the controversial and unstable orders over which men fight and despair; it passes unaffected by them, as daily life. The more it is exploited, the more the elementary things which our lives have in common come to light. The more numerous, varied and simple the people are who appear as subjects of such random moments, the more effectively must what they have in common shine forth. In this unprejudiced and exploratory type of representation we cannot but see to what an extent – below the surface conflicts – the differences between men's ways of life and forms of thought have already lessened. The strata of societies and their different ways of life have become inextricably mingled. There are no longer even exotic peoples. A century ago (in Mérimée for example), Corsicans or Spaniards were still exotic; today the term would be quite unsuitable for Pearl Buck's Chinese peasants. Beneath the conflicts, and also through them, an economic and cultural leveling process is taking place. It is still a long way to a common life of mankind on earth, but the goal begins to be visible. And it is most concretely visible now in the unprejudiced, precise, interior and exterior representation of the random moment in the lives of different people. So the complicated process of dissolution which led to fragmentation of the exterior action, to reflection of consciousness, and to stratification of time seems to be tending toward a very simple solution. Perhaps it will be too simple to please those who, despite all its dangers and catastrophes, admire and love our epoch for the sake of its abundance of life and the incomparable historical vantage point which it affords. But they are few in number, and probably they will not live to see much more than the first forewarnings of the approaching unification and simplification.

3 Virginia Woolf and English Culture*

Tony Inglis

Like Auerbach's chapter, this lecture, given at a conference in France on 'Virginia Woolf and the Bloomsbury Group' in 1974, is interestingly out of its time. In one sense, it is implicitly pursuing Auerbach's promotion of a Woolf who has not been granted her due, and who should be seen in relation to a European, not just an English literary tradition. Inglis is interested both in the kinds of arguments being made at the time by Woolf's own advocates, including those in France, and in the reluctance of the English critical establishment to take modernist writing seriously. He suggests that writers on Woolf should make their case for her in terms that will have more of an impact outside their own domain, and he also makes a succinct argument for modernist values that had up till then been more or less disregarded or misunderstood by critics uninterested in literary developments beyond the English-speaking world.

Professor Guiguet asked me to put forward, this morning, a 'reasonable anti-Bloomsbury view' as a gadfly's or Devil's Advocate element in your proceedings. I shan't in fact fulfil that expectation, though I hope not, in the end, to actually disappoint you. The 'reasonable anti-Bloomsbury view' came to seem, as I thought about it, both redundant and too easy. Too easy, because the mandatory objections, the appeals to alternative centres of value in the common people, or in a Tolstoyan ethic, or in the Dark Gods, have already been forcefully made; little would be gained by repeating or diluting them. Redundant, because time, social change and full documentation have put Bloomsbury beyond any simple polemic; the impulse to strike and declare a balance between the enviable and the pitiable, the fruitful and the sterile, the profound and the trivial in those lives, works and activities yields now to a dominant sense of their distance and singularity in relation to our own world.

---

* First published here in English; the French version was published in Jean Guiguet (ed.), *Virginia Woolf et le groupe de Bloomsbury* (Paris: Union Générale d'Editions, 10/18 series, 1977), pp. 219–41.

Moreover, I judged – as I reread some of Virginia Woolf's novels and looked over the recent critical writing about her – that for a group particularly interested in Woolf, the fruitfully provocative challenge lay elsewhere, in the quite wide discrepancy between the implications of the best of the recent studies and the wary or downright dismissive assessments that embody representative English opinion.[1] Much information, tending to clear up misunderstandings and to scotch error about Virginia Woolf, has come to hand in recent years – *A Writer's Diary* (1953) and Professor Bell's biography (1972) are the obvious landmarks. Comprehensive studies such as Professor Guiguet's, and special enquiries such as Professor Rosenbaum's into the influence of Moore, and Dr McLaurin's into repetition and rhythm, establish particular areas of clarity and perspective.[2] Later writers, critics and philosophers – specifically Beckett, Robbe-Grillet and Merleau-Ponty – in modifying our sensibilities, criteria and expectations, have altered our sense of what is significant in the recent past. This combination of scholarly work, critical thought and historical circumstance necessitates a revaluation of Virginia Woolf's novels and of her place in English literature. Yet, of the people who have brought this situation about, some are apparently still too much torn between conflicting sets of values to follow out the implications of their work and others have apparently lacked the audacity (in its double senses of 'the courage' and 'the cheek') to press their conclusions home. Towards those consummations I shall try to tempt and encourage you this morning.

Let us first review the traditional objections to Virginia Woolf's writing and see what modern criticism has done with them – have they faded away from sheer ineptitude, collapsed before argument, or merged into a higher synthesis? or do they lurk in the bloodstream of later critics, poisoning or confusing their enquiries?

Three of the old objections arose from neglectful misreadings – a negligence partly sanctioned by the 'affluent Bohemianism' of Bloosmbury and by Virginia Woolf's own early fluent, journalising and popularising attacks on Wells and Bennett, which put into currency an account of her aims and methods much less than adequate to the complexity and force of her best novels. I am thinking of the charges that her work lacks social range and concern – that it is undermined by a false poeticism – and that through lack of belief, 'a lax and vitiating monism' brings her fiction merely to accept dissolution, to surrender to the process of time, and to relinquish meaning in personal life.[3]

The charge of lack of social range, in any strict sense, can be refuted from any dozen pages of the fiction – the work-orientated lives, the poor, the mad, the blinkered-but-successful appear again and again: in the poor old woman by Regent's Park Tube Station and in the particularity of the lighthouse-keeper's boy's tuberculous hip, as much as in Septimus

Warren Smith and Mr Ramsay. An overt social-historical impulse (realised, admittedly, with varying degrees of engagement and wit) underlies *Orlando, The Years* and *Between the Acts*, and the evidence of social concern in the novels is amply borne out by the essays and by what we now know of the life. In so many ways Virginia Woolf was positively and advantageously the daughter of Leslie Stephen, the Victorian Liberal.

In dealing with the charges of false lyricism, we must distinguish between local and less damaging objections such as Fry's,[4] which may accurately diagnose particular failures or excesses even if we suspect Fry to be judging by a comparatively conservative norm of prose acceptability, and Miss Bradbrook's graver accusation that Virginia Woolf dealt *only* in sensitively recorded moments, 'epigrammatic metaphors' which fatigue the reader without significant selection and cohesion.[5] That more damaging charge has been fully dealt with by the many demonstrations (from Blackstone to McLaurin) of the tightness, coherent interrelation and felicity of Virginia Woolf's imagery and of her consistently dramatic rendering of percept and attitude together, of the intense *transitivity* of her writing.

Again, both before and after Savage's time, pondered reading and critical accounts tended to show that Woolf's novels are better read as weapons against flux than as inert surrenders to it. From Empson onward it had been possible to take for granted Woolf's *dynamic* use of the stream-of-consciousness convention; Daiches and Auerbach had shown how she *used* reverie, rather than simply reducing experience to reverie.[6] Savage's impatient and reductive accounts (especially of *Between the Acts*) are coloured and unduly sharpened by an anxiety over indeterminacy and the lack of absolutes that, a generation later, has been substantially overcome – we swim in the waves of flux instead of drowning in them.[7] But Auerbach's cultural placing of Virginia Woolf, in his closing pages, reminds us that even in the late 1940s it was possible to outgrow Savage's attitudes.

Two of the older objections to Virginia Woolf's fiction, however, sprang not from negligent reading but from adherence to conventions of fiction from which Woolf was consciously breaking away. The complaint that she cannot create 'character' seems to have subsided with the passage of time: our range of reference now includes the characters of Kafka and Camus, as well as those of Dickens and Thackeray, and it is difficult now to imagine a demand for memorable naturalism so strong that even Mr Ramsay hardly satisfied it. The complaint of lack of significant content, however, – a charge brought in their various ways by Leavis and his collaborators on the one hand and by Kettle and fellow Marxists on the other – is one that the general style and approach of Virginia Woolf's critics has done little to dispose of directly; and their common strategy of indirectness, their mode of sensitive commentary which entirely eschews paraphrase, has left this pre-modernist ideology flourishing in their own ranks to their partial confusion. To that aspect I shall return; for the moment, let me tackle the alleged lack of content itself. Here are representative objections: first, Kettle: 'What does Lily Briscoe's vision really amount to? In what sense is the episode in the boat between James and Mr Ramsay really a culmination of their earlier relationship?'[8] Second, Leavis's collaborator, Mellers: 'Shorn of the "original" technique, what Mrs Woolf has to say about the relationship between her characters, about the business of living, is both commonplace and sentimental.'[9]

Each critic, I suggest, makes his point only by denying or ignoring the nature of literature. Mellers makes the simpler error, the crude 'heresy of paraphrase' – 'shorn' of the particular verbal form Virginia Woolf evolved for her fiction, what Mellers sees is bound to be a naked and denatured travesty of the experience of the faithful and attentive reader. Kettle, as hungry for content and belief as Savage, whom he quotes with approval, goes further. The demand for assessment 'really' of a vision, and the enquiry about a 'real' relationship, where the relationship required to be *real* is a culmination of – perhaps only a transient moment of rest *within* – a complex psychological pattern, seems wholly inappropriate to the *kind* of testing and verification that the nature of the subject-matter permits. Kettle's reservations about Woolf's writing spring from a Philistine populism common in British Marxist criticism in the 1930s and 1940s: as a corrective I quote Walter Benjamin's comments on Kafka's modernist style. Everything, however surprising, says Benjamin, is said 'casually and with the implication that he must really have known it all along . . . as though nothing new was being imparted, as though the hero was just being subtly invited to recall to mind something that he has forgotten. . . . What has been forgotten is never something purely individual'.[10] I offer that first as an instance of sensitivity to the literary object – in this case Kafka's tone – taking precedence over the prompt ideological interpretation: second, as a valuable reminder of a whole mode of modernist profound indirectness in which Virginia Woolf (with Eliot and Golding, I think) participates – a mode which tends to escape from the categories appropriate to nineteenth-century art, and in which significance is not at all confined to the overt or the statable.

Against Mellers, and against this aspect of the *Scrutiny* rejection of Virginia Woolf's later work, one could usefully invoke Leavis's own sensitive and compelling analyses in defence of twentieth-century poetry wrongly accused, like our novelist, of incoherence, arbitrariness and vatic obscurity – the essays on *Mauberley*, on Hardy's poems of 1912, and the series on *Four Quartets*[11] – criticism notable for its generous, subtle and attentive reading, which allows meaning to be constituted as much by rhythm and juxtaposition as by the more paraphrasable elements of the poems. Again, D. W. Harding writes of Shelley: 'The contrasting and sometimes barely consistent ideas in these stanzas seem to have reached expression partly through verbal association that might be called accidental, were it not that they evidently gave openings for important variants of idea and attitude to emerge.'[12] Clearly the critical theory and practice of the group to which Mellers belonged could potentially have done ample justice to Virginia Woolf's style. More will be said later about the origin of their collective blindness. For the moment, let it suffice to have established that the particularly naive positivism of 'shorn of the "original" technique' is an isolated lapse, not a feature of *Scrutiny* critical method.

Similar demands for content and explicitness are made in our time. In *Tradition and Dream*, Walter Allen, discussing this novelist who renders the experience of flux, observes: 'her theme is a constant one, the search for a pattern of meaning in the flux of myriad impressions': the difficulty of life is reduced to the difficulty of a puzzle.[13] And the belief that there are puzzles with solutions is a fundamental blemish on Mr Leaska's book on *To the Lighthouse*;[14] he gives us a whole chapter of evaluative character-sketches, and one of his avowed motives, throughout his computer-analysis of the novel's style, is his wish to prove by mathematics a more negative account of Mrs Ramsay's character and significance than had previously been current.

In *Between the Acts*, the Revd. G. W. Streatfield falls into vacuity when he asks 'what meaning, or message, this pageant was meant to convey?' Does Virginia Woolf write a fiction of ethical assessment, of message, at all? Surely not – though in so far as ethical themes and judgements are adumbrated, I agree with Mr Leaska in finding them even-handed. But they are incidental, not central to the novel, as such judgements are in George Eliot or in early Henry James.

In following that positivist ethicising error into Allen's and Leaska's work, I have strayed across a borderline I had hitherto respected, between Woolf criticism of the reception period, the public responses down to the 1940s and early 1950s, and the contemporary period of Woolf criticism inaugurated by the publication of *A Writer's Diary* in 1953. Twenty and thirty years after a writer's death, we might expect the currents of opinion either to vindicate rather decisively the writer's lifetime reputation – as has happened, surely, with Yeats, Joyce and Lawrence – or to lead to a collapse, a general recognition of a lack of sustained first-order interest – of the sort that has overtaken Bennett and Wells and Galsworthy. Yet this resolution seems not to have happened with Virginia Woolf. There is still that curious air of embarrassment about her general reputation. '*Manquée*' said a distinguished English critic the other day when I mentioned her name, and her 'essentially minor stature' is still a commonplace in English textbooks – although, at the same time, no English novelist of the early twentieth century, except Forster and Lawrence, seems to be more highly regarded. My quotations from Kettle, Allen and the others do not fairly represent their central accounts of Woolf's work, which are much more sympathetic and apposite than could be gathered from my remarks today, but they do represent the pointedly negative element, the emphasis on limitation, which it seems *de rigueur* to bring in. One finds neither full and generous appreciation, nor cool diagnostic discussion.

This uncertainty about Virginia Woolf's critical reputation, this sense that misunderstanding and errors have been amply corrected but not dispelled, may stem from features of the fiction itself, or from very wide-ranging cultural factors which I'll take up in my next section: but some of the responsibility must rest with those who have written on Virginia Woolf. Woolf criticism, more than any other comparable body of discussion of a single author that I know, seems to have gone in for commentary and interpretation rather than for discussion aimed at changing received opinion – whether about the author's overall standing, or about the relative merits and significance of particular works. Of course, that tendency is a response to something in the novels. In his 1931 essay, Empson jocularly, and with a shade of irony, yearned for annotated and indexed versions of the novels, and several recent writers have seemed to offer just that. Conceptualisation and evaluative argument do seem particularly heavy-footed and inappropriate to the experience of reading this body of fiction. But they have been too completely neglected, and recent Woolf criticism (especially the weaker American offerings) shows the beginnings of an opaque and frigid cult. The academicisation of culture has had its effect as well: just as the overall reputation has stagnated, so *To the Lighthouse* has moved discreetly from being the one novel the pro- and anti-Woolf factions could agree to find merit in, to being the received classic which it's appropriate to use as *corpus vile* or test-bed for trying out a new critical method.

Putting it like that brings us back to Mr Leaska, whose chief motive I think I have summarily stated. His main interest is in the enterprise of establishing and analysing the objective, statistically-explorable evidence for consistent stylistic differentiation among the characters of *To the Lighthouse*. Even in his chosen field, there seem to be substantial faults. The method of choosing the samples leaves unchecked the possibility of variation within a character's style in the course of the novel, and in rigidly but subjectively assigning particular sentences (or parts of sentences) to particular characters, Mr Leaska has left himself open to charges of circular argument when conclusions are drawn at the other end. But the general attitudes and assumptions to which Mr Leaska turns are more to the point this morning. He is intolerant of ambiguity, however rich; he finds openness and uncertainty a source of distress, even in such classic cases as the Fourth Book of *Gulliver's Travels* and *The Turn of the Screw* (pp. 21, 38). In the closing pages, he facilely explains the multiple-consciousness novel as the offspring of an age of social and ethical individualism, diversity, and relativity which he deplores as 'cultural débâcle'.[15] He ethicises relentlessly, regarding even Mr Ramsay's inability to understand Cam's inability to understand the compass as evidence of Ramsay's 'sincere and uncompromising dedication to fact' (p. 122), whereas it's plainly – I take it – a point about the configuration and boundaries of the self, the at times grating and unfathomable otherness of our nearest and dearest. 'Real harmony could not be achieved until after Mrs Ramsay's death', he writes (p. 123), inadmissibly generalising the partial and temporary reconciliations of the end of the novel, as though (forsooth!) harmony is achieved when the imagery of battle gives way to that of drowning. Mr Leaska has embedded his worthwhile and quite substantial piece of work in a context of perpetuated critical misunderstanding.

If Mr Leaska's overemphasis on ethical evaluation in fiction represents one kind of disappointment in Virginia Woolf criticism, Mr McLaurin's recent book disappoints in another way. McLaurin is obviously a much more sensitive and sophisticated Woolf-reader than Leaska: he makes valuable connections outwards towards Proust, Cervantes, Lawrence and Sartre: his organising theme, the concepts of repetition and of rhythm, is apt and powerful. He raises his hat to Robbe-Grillet, and he uses Kierkegaard well. Again and again he seems on the point of expanding his precise and wide-ranging operations into a general onslaught on current opinion: but he avoids this body of issues, confining himself in a deadpan way to making intellectual connections and showing correspondences. If Leaska generalised too confidently and beyond his data, McLaurin seems trapped in detail. He too ends with a prolonged application of his ideas to a reading of *To the Lighthouse*, but he doesn't take any argued or explicit stand on the question of the novel's claim to be an apex; it seems to have been chosen through tradition and by the internal logic of the enquiry. That traditional central place of *To the Lighthouse*, going back through Auerbach to the very early criticism has led, one feels, to a 'screech as of brains wrung dry' in the multiplicity of gratuitous minor critiques, in the successive modes of Christian transcendentalism (c. 1950), existentialism and developmental psychology (c. 1960) and now, since 1970, a tentative *nouvelle critique*. It is too long since anyone tried explicitly either to vindicate or to dislodge it from its position.

The unargued, and perhaps delusory, aspect of that apparent critical consensus brings to our attention the neglect of internal evaluative questions in Woolf criticism. Cases are made for the last two novels –

and I believe at least one member of our *colloque* feels that there is much to be said for the first two novels also – but the key questions, 'You mean this is a *finer achievement*, as well as tightly organised, exploratory, interestingly new, and the rest?' seem in the main to be tactfully left unanswered. Unargued, at least, for here of all places we cannot forget Professor Guiguet's statement that *The Waves* 'is unquestionably Virginia Woolf's masterpiece': but he continues: 'the hostile comments, the objections called forth by *The Waves* . . . have nothing to do with literary criticism properly speaking, but are a matter of temperament or metaphysical attitude. A believer can only condemn an assertion of disbelief, a materialist an assertion of idealism'.[16] I think Professor Guiguet falters here at a crucial point in his case, a point where he could with advantage break out of relativism. If *The Waves* doesn't compel imaginative assent powerfully enough to overcome, in the mode proper to a work of literature, formal intellectual disagreement, then it's less than a masterpiece: to earn that honorific title it has to be the vehicle of a belief comprehensive and subtle enough to compel a cheer even from the ranks of Tuscany. I think it is, and I'm sorry that Professor Guiguet's case for the novel hasn't been argued in the evaluative and provocative terms which would compel adverse critics to defend their dismissal. The case for *The Waves* would have to be put in a wider context than the novel itself – especially by comparison with *Four Quartets*, that other modern English work that profoundly uses reverie and echo and multi-vocal monologue to embrace alternative significances and possible-other-cases: the mode of both works, one might want to say, achieved more than the ventriloquism of Auden or the vigorous cerebral structuring of Beckett. All this seems to me a necessary mode of discussion from which recent critics have been debarred by a dominant academic caution.

Lest I seem to be pressing an inappropriate demand for critical judgement and polemic edge from studies that set out more modestly to offer commentary or monographic exposition, let me show briefly some of the stresses and oddities in a book which is undoubtedly literary-critical in emphasis and addressed to the intelligent common reader rather than to the narrower academic public – Mr Moody's comprehensive and trenchant short study, which does pursue questions of value and which ranges more widely in the *œuvre* and the culture than Leaska does. It offers much valuable discussion, much stimulus to thoughtful disagreement. Mr Moody makes the strongest case possible for Woolf within the dominant English critical tradition and idiom of his generation – a tradition and idiom whose demands for reference and judgement trap him into a more sophisticated form of Leaska's error. *Jacob's Room*, we gather, is primarily about Edwardian civilisation and the Great War, rather than experiment in form and ontology; *Mrs Dalloway* ('a minor and imperfect work') is essentially satiric, with Clarissa more truly parallel to Sir William Bradshaw than to Septimus Warren Smith – for Moody, 'The crucial flaw in the dramatisation of Clarissa Dalloway's "death of the soul" in that the connexions with her "doubles", Bradshaw and Septimus, are not brought home directly enough really to touch her own complacent image of herself'.[17] There Moody actually concedes that the novel is not what he takes it to be, and describes its divergence from his preferred emphasis as a 'crucial flaw'. Elsewhere, noting correctly that Woolf writes 'around the absence of a fulfilled natural vitality, he assumes that she must therefore be simply and directly yearning for the missing quality and, astonishingly, takes Mrs Manresa in *Between the Acts* to be offered as a genuinely 'wild child of nature' fulfilling that need.[18] Again, to show the inadequacy of Woolf's notion of fulfilment, he invites us to compare the famous rhythmic passage from the first chapter of *The Rainbow* – a standard citation among vitalist readers of Lawrence who overlook the critical, undermining nuances in Lawrence's prose. Running thus with the hare and hunting with the hounds, Moody praises Woolf where she is least herself. He makes her more Lawrentian and more prescriptive than she ever was in her art. His admiration for the rough unfinished aspect of the pageant in *Between the Acts* comes out more strongly than any feeling for Woolf's hard-worked achieved artistry, and he writes with enthusiasm, 'There is none of the hysteria of the isolated and alienated literary mind' (p. 85), dismissing so casually so many European writers between Baudelaire and Beckett. Moody's final emphasis is on the novelist's working towards the projection of an 'ordered wholeness' – still, in effect, the positivist ethical account. Such are the *obiter dicta* of a critic whose enterprise, in relation to Virginia Woolf, is quite close to my aim this morning.

The distinctive weakness, then, of current Woolf criticism is this: the various crude objections already rehearsed – items, mostly, on Monsieur Robbe-Grillet's list of 'Outdated Notions' these twenty years, already amply refutable by attentive reading of the novels, and amply refuted *ad hoc* by scholarly and critical demonstration – don't seem in fact to have been killed. People still say, 'the very best sort of minor writer', or '*manquée*'; or they reject on essentially irrelevant traditionalist grounds; or they express enthusiasms which – even when they support them with the care and judiciousness and learning of Professor Guiguet – they concede too soon to depend ultimately on deep personal choices; or like Mr Moody they try to acknowledge and define the particular qualities, but at the same time to satisfy the traditional fictional values of reference and maturity. No one has backed Woolf hard enough, or with enough conviction, to carry through the review of modern English literary history that full acceptance of the implications of current claims for her entails. I think I have shown that the difficulty does not lie in any cogency in the negative arguments themselves. Something so widespread must be a response to a climate and a context rather than a matter of limitation or idiosyncrasy in particular critics. Are we not in contact with that central feature of modern English literary and artistic life, the imperfect assimilation of – and indeed the strong resistance to – modernist art in English culture? That resistance, in its origin and in its primary forms, has by no means been all loss – it was up to a point a source of creative tension for Lawrence, Eliot, Joyce, and Woolf herself; but some of its secondary forms, notably the long deafness of English philosophers to the Continent and a reactionary vulgarisation in literary criticism in recent years, have been parochial and restrictive in their effects.

The cultural history I have in mind is roughly this: modernist art is above all epistemological both in the shift in sensibility in which it originates and in the tendencies which it displays and most characteristically develops. Embodying a reaction against the empirical certainties and linear logic of the mid-nineteenth century, from the standpoint of the very unstable blend of freedom and limitation, power and impotence characteristic of twentieth-century experience, it replaces God by a felt absence, syntax by discrete images, historical progress by repetition, the proscenium arch by the open space amid the audience, and so on. Superficially individualist, it frees the individual from the social frameworks of country, church and family (those 'nets' from which Stephen Dedalus thought to escape) – but the individuality thus achieved, quite unlike nineteenth-century affirmative liberal selfhood, is deeply collective, apprehended through the body, the unconscious and the mediating functions of language and myth in relation to consciousness. (This is of course a compressed account of central tendencies as we can now see them – it does not attempt to account for paradoxical cross-currents such as the social thought of Hulme, Eliot and Lawrence.) The seed of such a movement inevitably fell on thin soil in England, where its exponents were mostly foreigners (Pound and Eliot), Irishmen (Yeats and Joyce) or exiles (Lawrence). England had been so fully and successfully the generative centre of that liberal empiricist culture of ethics and history that it even rode out the shock of the Great War as well as those of relativity and psychoanalysis. Forster offered his diagnosis of the Wilcoxes before 1914, but the tone and style of British public life, in the 1930s and even in the 1940s, retained much of the flavour and confidence of Victorian and Edwardian Liberalism. The native tradition of opposition was a flamboyant aestheticism with its roots in the 1890s: the Sitwells were its archpriests, Bloomsbury certainly one of its congregations.

While the intellectual élites just below the level of the creative geniuses themselves were absorbing and winnowing the new art – a process necessarily extended over two or three decades – two new waves of cultural change swept ashore.

One, reflecting England's position off the European mainland and in many senses part-way to America, was the impact of the new American culture of film, cars and the associated consumer values: an impact which, for the conscious Englishmen, threw into relief the cultural virtues of the nineteenth-century middle class and of the still older rural culture – the one ethical and humane, the other customary, organic and face-to-face, despite hierarchical and even oppressive elements in both.[19]

Second, the perhaps inevitable and appropriate return of some social concern and content into modernist art (after the initial battles of idiom and sensibility had been won) was immensely hastened and intensified by the economic troubles of the 1930s, the rise of the Continental dictatorships and the politicisation of culture; in the 1930s, after all, the author of *Ash Wednesday* wrote *Murder in the Cathedral*, the author of *The Waves* wrote *The Years*, and the author of *New Bearings in English Poetry* turned his sensitivity, energy and polemic power away from the (admittedly now second-order) challenges of the new towards the great fiction of the previous century. It is easy to understand and to credit the impulses behind these moves, and not easy to argue firmly that anyone was mistaken: but the upshot was to leave unscotched, and perhaps apparently vindicated by the most distinguished minds of the age, a line of conventional, nostalgic and backward-looking cultural criticism, in the camps of Eliot and of Leavis alike.

The tendency in those years, then, was to assimilate, tone down, defuse continental and internal modernist forces. After 1945, the social and educational revolution and the renewed dislike of America (which had plainly won the 1939–45 war so much more thoroughly than we had) bred a new school of plain-man, positivist novelists and critics ranging from such respectable upper-middle-brow novelists as Angus Wilson, down to the deft and pointed Philistinism of *Lucky Jim*, and perhaps represented on a middle level by Donald Davie's recent attempt at marshalling Hardy, Lawrence and Graves into a line of native English poets unsullied by the Franco-American mannerism of Eliot and the rest.[20] The strongest counter-tradition to this shift in literary opinion was the Leavisite one; but Leavis's emphasis on major continuities with the past remained the mainspring of his visible and current activities, and his influence became more merely conservative, moralistic, and positivist as it was diluted downwards through the universities and schools.

In that cultural context the scandal of the native English modernists has been dealt with. With Lawrence, the usual move is to define modernism as essentially a matter of *form* and thus to detach Lawrence – and at the same time to transpose him into a paraphrasable sexual and social 'prophet'. The drawing of morals (or, with American critics, the superimposition of patterns of ethical and symbolic antitheses) can then proceed undisturbed. The same procedure – denature, normalise, trivialise – is applied to Virginia Woolf – as English as Lawrence, and writing like him out of that liberated turn-of-the-century radical culture of Bergson and Nietzsche. If she is not declared *manquée* for not writing like George Eliot, she is discussed either in terms of poeticism (and thus reduced by Lord David Cecil to a tremulous female aesthete) or anecdotally as a member of that grotesque and disreputable bunch of columbines and pathics 'Bloomsbury'. Either way, the sturdy Englishman is safe. Hence Mr Moody's rejection of Anguish – though his mentioning it at all, and the whole painfully straddled posture which he maintains in his book, may mark an impending change. This is the context in which it is high time Woolf's sympathetic, learned and hard-working critics – perhaps above all Messrs Guiguet, Moody and McLaurin – should throw away their long-maintained discretion and draw, publicly and overtly, the conclusions they seem to be withholding. Only by their doing so can the wider, external issues surrounding Woolf's stature and reputation be resolved.

Those issues embrace, first, a direct attack on the question of Woolf's standing as a novelist – the question that evokes so many subtly dismissive answers. Of course there is no international league table in such matters, and the wider the range of comparisons one seeks to make, the less apposite the results will be. Let me risk a few tentative comparisons, none the less. Among the English novelists of her own time – Hardy, Forster, Bennett, Lawrence – Woolf clearly stands high. Only Lawrence – who certainly has his unevennesses and *longueurs* – can stand comparison for the scope and range of the embodied life in the fiction and for originality and distinctiveness of presentation and articulation. Equally, it seems entirely in place to discuss her along with James and Conrad, whose concerns overlap substantially with hers. She does not of course share in whatever merit one grants to the *summa*-writing modern novelists, Joyce and Proust, and there is a 'world class' (Dickens, Tolstoy, Dostoevsky) who set standards by which Woolf dwindles – to rise again when she is placed alongside the higher commercial product offered by Huxley, Compton-Burnett, Murdoch and the like, who have so much less to offer. The body of finished work, the sense of a substantial literary personality creatively embodying being and milieu, place her close to Jane Austen and George Eliot – and if George Eliot displays a greater density, enjoys a more direct relationship to an art-form that had not in her time become problematic, the discrepancy seems to lie less in quality than in sensibility and in historical position, a contrast analogous to that between Marvell and Donne.

Second, demolition of the fence round Woolf's reputation, her acknowlegement among the major, mainstream modern English novelists and the consequent revision in readings and contexts of her work, should compel an overdue adjustment and clarification in the relation between the ethical and the epistemological aspects of the novel. The reigning view in England – that the ethical is primary, the epistemological almost nowhere – flatters the plain-man traditionalists in British culture, and keeps the world of appearances safe and comfortable at the cost of undervaluing Woolf, Sartre and Kafka, partly misreading Lawrence, and being not good at dealing with Beckett or indeed with French literature since Proust. Woolf, properly read, could help us towards the more balanced view that these two aspects of the novel vitally interpenetrate one another. Whether the novelist begins, like Lawrence, from a (Nietzschean) moral interest and is led into his later investigations of Being, or whether she begins like Woolf with a Humean concern with identity, solitude, memory and other minds and is led into ethically-charged particularities of experience, both aspects are typically present – in Woolf and Lawrence, as in James and Sartre. Even in Austen and Eliot, writers who seem to move confidently enough among social surfaces, the ethical content turns on the discovery and definition of the self through the felt realisation of the autonomy of other centres of being (Miss Bates, Grandcourt). This aspect, worked out in the detail of the novelist's prose, leads the novels out of the particularism or the banality of their specific ethical positions. This is too large a topic to be fruitfully taken further today, but clearly it is bound up with the question of Woolf's mode and stature: and it has not been explored.[21]

Third, recognition and clarification of the scale of Woolf's achievement should stimulate and enrich critical and historical thought about modernism in English and in other literatures. Not only does she offer another instance of the problem of sustaining career and *œuvre* in the face of the rigour and extremism of modernist concerns and techniques – a case to consider beside Kafka's reluctance to publish his fiction and Sartre's to finish his, the striking break in the poetic career of Eliot, the pressures towards assimilating language and experience into a *summa* (like Proust and Joyce) or refining them into a silence like Beckett. Like Yeats and Thomas Mann, Woolf is one of the converts to modernism, whose substantial juvenilia in previous or transitional modes show continuities going back, in her case, to James and even Pater; and, like Yeats's, her converted style continues to grow and develop instead of hardening, as did, say, Pound's. Moreover the date of her conversion – just after the intensely polemic, promotional period associated especially with Pound, yet before the time when the epigoni Auden and Beckett deploy various stylisations to resolve (and in part to evade) modernist technical and substantive dilemmas – puts her at a distinctive stage in the movement, less frenetically doctrinaire than her predecessors, more fruitfully direct than her immediate successors. In these, and doubtless in other connections, taking Virginia Woolf seriously will illuminate problems of continuity and tradition in modernism – a prerequisite both to clear thinking about our current relation to inter-war modernist art and to the development of a properly aware successor culture to modernism.

For all these reasons I urge Virginia Woolf's critics to abandon their excessive modesty, to argue a case at once explicative, evaluative and polemic, and to take a full part in the debates which will follow from their doing so. Some steps in this direction have been taken by Mr McLaurin, and (in contrast to the general drift of Woolf criticism in America) by Messrs Freedman and McConnell,[22] whom I have not managed to include in my survey. In addressing my incitements to a French audience, I take it for granted that any work along these lines that I may have the good fortune to stimulate can only benefit from the triple presence in your culture of the *nouvelle critique*, with its emphasis on the encoded and the spatial in fiction, the *nouveau roman*, with its anti-naturalist stress on the made, artefactual quality of the novel, and the phenomenological approachs which seem so much more accepted in your intellectual life than in ours across the Channel.

Notes

1. For instance: 'a minor talent . . . who upheld aesthetic and spiritual values in a brutal, materialistic age' (F. W. BRADBROOK, *The Pelican Guide to English Literature*, vol. 7, *The Modern Age* (Harmondsworth: Penguin, 1961), p. 268; 'the pattern of art as artifice . . . limits the ultimate scale of her modernism' (MALCOLM BRADBURY, *The Sphere History of Literature in the English Language*, vol. 7, *The Twentieth Century* (London: Jonathan Cape, 1979), p. 202.

2. JEAN GUIGUET, *Virginia Woolf et son œuvre*: (1962), trans. Jean Stewart (*Virginia Woolf and her Works*, London: Hogarth Press, 1965); S. P. ROSENBAUM, 'The Philosophical Realism of Virginia Woolf', in Rosenbaum (ed.), *English Literature and British Philsophy: A Collection of Essays* (Chicago: University of Chicago Press, 1971), pp. 316–56; ALLEN McLAURIN, *Virginia Woolf: The Echoes Enslaved* (Cambridge: Cambridge University Press, 1973).

3. D. S. SAVAGE, *The Withered Branch* (London: Eyre and Spottiswoode, 1950), p. 96.

4. 'She is so splendid as soon as a character is involved . . . but when she tries to give her impression of inanimate objects, she exaggerates, she underlines, she poeticizes just a little bit', in Denys Sutton (ed.), *Letters of Roger Fry* (London: Chatto and Windus, 1972), vol. II, p. 598.

5. M. C. BRADBROOK, 'Notes on the Style of Mrs Woolf', *Scrutiny* (1932), 36. By 1970 Miss Bradbrook no longer agreed with her earlier judgement.

6. WILLIAM EMPSON in Edgell Rickword (ed.), *Scrutinies by Various Writers*, II (London: Wishart, 1930), especially p. 211; DAVID DAICHES, *Virginia Woolf* (Norfolk, CT: New Directions, 1942); ERICH AUERBACH, *Mimesis: The Representation of Reality in Western Literature* (1946), trans. Willard R. Trask (Princeton: Princeton University Press, 1953).

7. Compare the ending to Part II of MAURICE MERLEAU-PONTY, *The Phenomenology of Perception* (1945; trans. London: Routledge & Kegan Paul, 1962):

> Being established in my life, buttressed by my thinking nature, fastened down in this transcendental field which was opened for me by my first perception, and in which all absence is merely the obverse of a presence, all silence a modality of the being of sound, I enjoy a sort of ubiquity and theoretical eternity, I feel destined to move in a flow of endless life, neither the beginning nor the end of which I can experience in thought, since it is my living self who thinks of them, and since thus my life always forestalls and survives itself. Yet this same thinking nature which produces in me a superabundance of being opens the world to me through a perspective, along with which there comes to me the feeling of my contingency, the dread of being outstripped, so that, although I do not manage to encompass my death in thought, I nevertheless live in an atmosphere of death in general, and there is a kind of essence of death always on the horizon of my thinking.

8. ARNOLD KETTLE, *An Introduction to the English Novel*, vol. 2, *Henry James to the Present Day* (London: Hutchinson, 1953), p. 105.

9. W. H. MELLERS, 'Mrs Woolf and Life', *Scrutiny*, VI (1937), 72.

10. WALTER BENJAMIN, 'Franz Kafka: On the Tenth Anniversary of his Death', in *Illuminations*, trans. (1955; Henry Zohn repr London: Jonathan Cape, 1970), p. 131.

11. On Pound in *New Bearings in English Poetry* (1932; new edn London: Chatto and Windus, 1950); on Hardy in 'Reality and Sincerity: Notes on the Analysis of Poetry', *Scrutiny*, XIX (1952), 90–8; on Eliot's *Four Quartets* in *Scrutiny*, XI (1942–43), 60–71, 261–7, and *English Literature in our Time and the University: A Study of the Contemporary Situation* (London: Chatto and Windus, 1969), Chapter 4.

12. D. W. HARDING, 'The Hinterland of Thought', in *Experience into Words: Essays on Poetry* (London: Chatto and Windus, 1963), p. 190.

13. WALTER ALLEN, *Tradition and Dream: The English and American Novel from the Twenties to our Time* (London: Phoenix, 1964), p. 18.

14. MITCHELL LEASKA, *Virginia Woolf's Lighthouse: A Study in Critical Method* (London: Hogarth Press, 1970).

15. For a diametrically opposed and more suggestive account, see LUCIEN GOLDMANN, *Pour une sociologie du roman* (1964; repr Paris: Gallimard, 1970), pp. 49–52.

16. GUIGUET (see note 3), p. 297.

17. A. D. MOODY, *Virginia Woolf* (Edinburgh: Oliver and Boyd, 1963), pp. 14–17, 18, 27.

18. MOODY, p. 7; compare *Between the Acts* (New York: Harcourt Brace Jovanovich, 1941), p. 41: 'So with blow after blow, with champagne and ogling, she staked out her claim to be a wild child of nature.'

19. 'France is really the chief hope of any resistance to America; we have already given in . . . Only we keep going individually – it's only the general life that's so appalling – we can have no public art, only private ones, like painting and

writing, and even painting is almost too public', *Letters of Roger Fry* (see note 4), vol. II, p. 631.

20. DONALD DAVIE, *Thomas Hardy and British Poetry* (London: Routledge & Kegan Paul, 1973).

21. McLaurin makes an apt small point about *Orlando*'s being Sartrean in its rejection of 'picture-gallery history', but he does not proceed to connect Clarissa and Rhoda with Roquentin.

22. RALPH FREEDMAN, *The Lyrical Novel: Studies in Herman Hesse, André Gide and Virginia Woolf* (Princeton: Princeton University Press, 1963; FRANK D. McCONNELL, '"Death among the Apple Trees": *The Waves* and the World of Things', *Bucknell Review* XVI (1968), repr *Virginia Woolf: A Collection of Critical Essays*, ed. Claire Sprague (Englewood Cliffs, NJ: Prentice-Hall, 1971), pp. 117–29.

4 Mirrors and Fragments*

FRANÇOISE DEFROMONT

Moving between the texts of Woolf's novels and those of her life-history, this part of Defromont's book looks at moments which evoke the beginnings of an identity fragmented even in its formation as potentially whole. Her shaping myth is that of psychoanalysis, filtered in particular through the work of Jacques Lacan. For Defromont, the splittings of something that was never complete continue to replay themselves in the shocks and jolts of characters' lives, but they also impinge upon the narrative mode of modern texts that are 'in pieces' in a way comparable to the bodily image or the self that seeks to make itself whole. Like Abel and Jacobus, Defromont is especially interested in how the figure of the mother functions in Woolf's writing to focus the ambiguities and impasses of a female predicament, in particular those that concern the would-be woman artist.

Mirrors

That little girl described in *Moments of Being* seeking her image in the mirror – is she immortalised in a 'real' memory? Are we really dealing with a biographical fact? Every child has lived this symbolic moment; it corresponds to the questions: Who am I? Do I love myself? These questions, which Virginia Woolf never stopped asking herself, resonate throughout her *œuvre*. The mirror is not only a theme, it is the sore point where all the unconscious currents which run through the text meet up.

Coverings inundated with light; transparent flashes; shimmering liquid surfaces. Mirrors of all sorts. Woolf has the art of causing the flaming or translucid flash of windows daubed by sunlight to shine out, or else of grasping the fleeting light of the setting sun reflected on glasshouses (JR, p. 9).

---

* Translated by Rachel Bowlby from Defromont, *Virginia Woolf: Vers la maison de lumière* (Paris: Editions des femmes, 1985), pp. 107–29. Some modifications of syntax and sense were made to the original text, in consultation with the author, and I would like to thank Françoise Defromont for her enthusiastic collaboration. All ellipses occur in the original.

It could even be said that this writer uses the principle of the refraction of light as a true literary technique. This implies for example a kind of approach to characters by which they are not presented directly, in totality, but transparently, with many openings: the character is composed like a mosaic made of hundreds of tiny flashes.

In *Mrs Dalloway*, Septimus is sometimes captured across his thoughts, sometimes across the look of Rezia, his wife. Then he is momentarily glimpsed by a passing woman, and finally he is fleetingly spotted by Peter Walsh, another of the novel's characters who crosses his path without knowing him. Woolf circles around her character as though she had a camera in her hand, grasping him from multiple points of view, from the most intimate to the most external.

There is no definitive truth; each character has his/her own truth which the reader can surprise across the transparency of another look. The consciousness of each serves a kind of mirror-function, such that the personality of the other protagonists is picked up and reflected off again, continually, from one mirror to another.

This process is really staged, or acted, in *Between the Acts*, a novel in which mirrors of all sorts suddenly return to the astonished spectators their own faces, their own attitudes, but fragmented, which is why the spectators cannot be 'whole' (BA, p. 135). There is no centralising entity, only a fleeting halt to the flux of life, and we are led to wonder whether the author's role is not merely to be there, in the manner of the actors of *Between the Acts*, behind the mirror, and to hold it tilted at a certain angle. So the omniscient author does not exist; the role of the producer is in some sense to promote the phenomenon of refraction between the two sides; one can only admire the brilliance of Virginia Woolf staging her own technique in this passage. The fragment of reality is captured in an endless to-and-fro; it brings about a trajectory which unfolds between stage and spectators, or author and reader, or again between writing and reading; this resembles the play between two mirrors, and one wonders where the point of departure should be situated. Who then serves as origin in the simultaneity of the reflected image? This double trajectory takes us back to those questions of identity and origin which could be summarised as 'Where do I come from?' – questions which point to the mother. The other side of the mirror is the house of light, round which are gathered the questions of identity and femininity. When the maternal presence is radiant, the mirror sends back a satisfying narcissistic image; this corresponds to what I will call the mirror of light. Mrs Ramsay receives the rich and mysterious light of the lighthouse, impregnates herself with it and becomes herself a reflecting surface: she absorbs the light to radiate it out again in her turn. The mirror is alive: 'It seemed to her like her own eyes meeting her own eyes' (TL, p. 61); there is complete coincidence and identity between the two luminous points; the mirror does not divert the image but restores it perfectly. Similarly, in *The Voyage Out*, the hotel and the lit-up house respond to one another like two mirrors facing each other, or rather, like two equal forces, both sources of light. Thus, when the mirror fulfils its function, the image of self seems to be whole and alive.

Transparency and opacity are linked to drives of life or death which go back to luminous sources or glazed windows. When they are transparent, they allow free circulation to light and are synonymous with life; when a gilded reflection is immobilised there, the translucid panes of the glasshouse change it into a house of light.

The transparency of the pane is associated with the movement of light; it is a window giving onto the intimacy of the self. The characters' 'preferred station' is to be by the window, whence they look out at the world: to go to the window is like the sign of an opening onto the soul and onto the profound thoughts of him or her who dreams, and whose interiority thus communicates with what is exterior.

Transparency of the window-pane and transparency of the water. The first lines of *Jacob's Room* are like an opening onto Betty Flanders' emotions, and the bay, the lighthouse and the waves which fall within her range of view are perceived across the transparency of her tears. The tears flow over the letter she is writing, mix into the ink as though to give to Woolf's writing the transparency of water.

Yet the smooth, clear surfaces are sometimes deceitful and suddenly become as opaque as a wall. Thus Sasha, with whom Orlando falls deeply in love, is stunning on the outside, but something in her sounds hollow, since her radiance is not internal; it resembles a wandering flame (O, p. 30). The source of Sasha's light is not from within, and there is no circulation between the inside and the outside, which is why we do not rediscover in her the radiance and warmth of the house of light.

The amorous meeting between the two characters occurs at the time of the Great Frost, which covered London with an immense blanket of ice, nothing other than an illusion of transparency. Its frozen surface is rigid like death, for it is as hard as steel. The frost is like the reverse of the fluid movement of the wave: the ice has forever petrified those who let themselves be caught in it. In the same way that the window sometimes turns into a prison behind which the character is shut in and present, powerless, at the procession of life, so, when transparency is transformed into quasi-metallic opacity, it becomes an unbreachable wall. The butterfly beating against the window is then found dead; in *Between the Acts*, the wall becomes the symbol of fallen civilisation, of war and despair. It is the total negation of transparency and light. Identity is then tied to death.

Splits and doublings

When the mirror ceases to be a luminous covering, its opacity provokes the ego's double; it no longer reflects the image. Physical descriptions may be followed by the immediate displacement of the projector towards a picture beside or above the person being described. The mirror-function goes off-beam, and we are confronted by two distinct images.

This displacement of the camera resembles a distanced commentary, or serves as counterpoint to physical description; this produces an effect of subjective division as though two people were cohabiting inside the same individual. In *Mrs Dalloway*, the masculine side of Lady Bruton's personality is expressed through the portrait of her father, and she is seized with the same immobility as the old general holding his parchment, in the picture which is behind her (MD, p. 116).

The symmetry of the gestures recalls a reflection caught in a mirror; yet there is a bursting apart of two opposed aspects of the character; the mirror no longer returns anything but an immobilised image in a mortal trance. In this context, the ego is split between masculine and feminine and distributed between the woman Lady Bruton is and her father, the man she bears within her. The same phenomenon can be observed in *Between the Acts*, where two pictures are hung up next to each other, as a mirror, in the empty dining room. (BA, p. 31).

The 'heart of the home' contains only 'void, silence'; we remember the early memory evoked in *Moments of Being*: the bowl was full and overflowing with happiness despite the double breakage in the centre. In *Between the Acts*, the space is immensely empty; it is the expression of the absolute absence of the maternal figure and the sign that identity has burst apart.

What is the meaning of the void and splitting associated with the mirror? We cannot fail to mention what Lacan calls the 'mirror stage'; the hypothesis is that at an early stage, the child has only a divided perception of its own body. At the moment when it recognises its image in the mirror, an important phase in the evolution of its personality has been crossed, for in its own perception its body is organised into a coherent whole; its identity has begun to be structured.

As far as Virginia Woolf's *œuvre* is concerned, we will be interested above all in how the text itself produces the effect of a division into pieces.

The novels raise the question of the mirror stage in an insistent way. The mirror is full when the maternal image is restored in its integrity; thus, in *To the Lighthouse*, the mother's radiance flows all through the text, even after her disappearance. And on the other hand, it would seem that the mother's absence forbids the text to go on to the mirror stage, to the extent that the image of the self is inevitably doubled, as can be seen to happen in *Between the Acts*, where the broken flashes remain disjointed. The furthest point of the splitting is inscribed in this last example, since the question of identity is no longer played out except between two inert and fixed pictures.

The void is situated at the heart of the text, just as it is hollowed out at the heart of identity; thus the theatrical performance improvised in the middle of the countryside by Miss La Trobe to tell the history of England in the form of sketches is also empty. Further, there is a decentring, for the action is displaced offstage where the cows constitute the only mobile element; there would be much to say on the subject of the cow as maternal figure. . . . The question of the centre remains in the air since, as one spectator says, it is what everyone present needs (BA, p. 127). We might imagine that the empty mirror figured by the decentring of the theatrical scene denoted some kind of narcissistic suffering, as if, ultimately, the gratifying reward of the image were failing to return to the subject looking at himself in the mirror. Empty mirror, mirror shattered by the disappearance of the maternal figure?

Is there an answer to the question of the self which the text raises over and over again? How is the question of identity unfolded? Something which might approximate to a response, or at least to a mooring which would prevent us from drifting, has to do with books. 'Books are the mirror of the soul. . . . So the mirror which reflects the sublime soul, also reflects the bored soul' (BA, p. 144). If we go back again to biography, we find an analogous trajectory when we once more see Virginia hidden away at the top in the children's room, with her books as the only remedy against the shackles of Victorian society.

Yet while there may be a unison between the soul and its image reflected in the mirror, the comma which separates the sentence into two parts marks a division. The split constantly inscribed touches essentially, it seems to me, on the problematic of sexual identity and could be translated like this: am I woman, am I man?

To bring together these various elements, we could say that because of the absence of the maternal figure, the ego does not succeed in organising itself into a coherent whole, or in constituting for itself a satisfactory narcissistic image. This ultimately provokes an uncertainty as to sexual identity. The biography, once again, enables us to elucidate these phenomena: we had picked up from the text an echo of the trauma provoked by the death of the mother; it is as though her absence had deprived the writer of a female figure to identify with. There remained the father-figure; and the father was committed to literature and dedicated to books.

The book-mirror is extremely important in the text; it gives the illusion of filling the void. In *Orlando*, the vast oak tree from which the character contemplates the sea, as from the top of a lighthouse, is also, as we know, a book. It is the original manuscript of the text Orlando writes over the centuries. The oak–book is central to the text as it is central to identity, in the place of the house of light.

So what happens? The search for a feminine figure with whom to identify goes via the house of light; the oak tree shifts the question in the direction of the phallic. The maternal figure then appears two-faced: the house of light, which is pure femininity, is doubled into its reverse side, the oak–book; thus the mother too appears phallic.

Even if the book apparently fills a void left in the centre, a gap is left open; text with holes, text whose centre of gravity is always retreating towards the edge, hollowing behind it a gulf which designates a space between the acts. . . .

A fleeting interval suspends the rhythmic movement by a violent, but almost invisible interruption: a muffled interlude designates the disappearance of the loved woman, or the missing scene which makes the void around her because this scene is unbearable.

The mirror shattered, or a horror scene

The mirror is shattered at the moment when there is a double breakage. Biographical echoes resonate through the writing; the two traumatic scenes inscribed in *Moments of Being*, rape and death, are rips which echo everywhere.

Two of the novels reflect the absolute void provoked by one or the other of these traumatic scenes, which take place exactly in the middle of *Between the Acts* and *The Waves*. In *To the Lighthouse*, on the other hand, despite the disappearance of Mrs Ramsay in the middle section of the text, something of the radiance of the house of light subsists.

At the centre of *The Waves*, there is – nothing; the death of Percival is inscribed without taking place, and this designates the utter horror of absence and death. The death of the mother is thus enclosed in the death of the brother; a doubled mourning which transforms the text into a coffin, a prison of death.

This icy novel, which does not manage to warm itself up again in the rays of sunlight retained and distanced by the barrier of the poetic interludes, should be read in conjunction with *To the Lighthouse* and *Between the Acts*. For the text really is inhabited by a central void, whereas in *To the Lighthouse*, in contrast, heat and light spread out in spite of the internal stripings of black and white. Between these two poles, *Between the Acts* is the text in which the meaning is revealed, for it contains the missing fragment which makes it possible to decode at closer quarters the problematic of void and shattering in Woolf's *œuvre*. This is the place

67

where that monstrous scene is completely unveiled, in all its horror; here it is:

> Giles, nicking his chair into its notch, turned too, in the other direction. He took the short cut by the fields to the Barn. This dry summer the path was hard as brick across the fields. This dry summer the path was strewn with stones. He kicked – a flinty yellow stone, a sharp stone, edged as if cut by a savage for an arrow. A barbaric stone; a pre-historic. Stone-kicking was a child's game. He remembered the rules. By the rules of the game, one stone, the same stone, must be kicked to the goal. Say a gate was a goal; to be reached in ten. The first kick was Manresa (lust). The second, Dodge (perversion). The third himself (coward). And the fourth and the fifth and all the others were the same.
>
> He reached it in ten. There, couched in the grass, curled in an olive-green ring, was a snake. Dead? No, choked with a toad in its mouth. The snake was unable to swallow; the toad was unable to die. A spasm made the ribs contract; blood oozed. It was birth the wrong way round – a monstrous inversion. So, raising his foot, he stamped on them. The mass crushed and slithered. The white canvas on his tennis shoes was bloodstained and sticky. But it was action. Action relieved him. He strode to the Barn, with blood on his shoes.
>
> (BA, p. 75)

This scene is fundamental, first because all the traumatic scenes of rape and death in Woolf's life intersect in a written symbolisation; and second, because we are present at the emergence of another element: the inversion of the value of the masculine in relation to the feminine – the solar mother, the mother of beauty, having swallowed the father's phallus and thus constituting herself as referent in its place. All these elements are joined and produce a scene in which reversals and breakages of all sorts are inscribed in all their brutality.

I am now going to attempt to unravel this scene step by step: all the questions relating to the body, identity and sexuality are there. It begins with Giles's turning round. Hidden scene: you have to turn to (not) see. The context is primitive, savage, barbaric (hardness of the stone, dryness). Giles, the seducer, the 'male', plays at getting the stone in the goal in ten shots. The first three shots are called lust, perversion and coward(ice).

The scene is sexual; the crossing of the threshold prefigures and heralds the monstrous coupling sheltered behind this preliminary moment: the snake can no more swallow than the frog can die. A scene both traumatic and primal: the fixation is complete, there is contraction and a mortal trance: it is the opposite of orgasm. Let us note in passing that some elements are already familiar to us: the rigidity of the trance points equally to Leonard's account of Virginia [in *Beginning Again*, discussed earlier in Defromont's book] and to the divided portrait of Lady Bruton.

The dénouement does not take place. Neither the snake nor the frog succeeds in disengaging itself from this horrible impasse. The arrest of all movement also makes us think of the petrification of the Great Frost: ice of the body, ice of *jouissance*.

This sexual act is unveiled as being a 'monstrous inversion': and there is indeed a reversal between the masculine and the feminine. It resembles a primal scene, in other words the coupling of the parents which takes place when the child's back is turned. All this thus seems doubly incomprehensible; and it really is pretty strange: who is acting the male, and who is acting the female? The snake, generally perceived as a phallic figure, is here in the position of absorbing the frog: it is thus transformed into a container, stomach or vagina. And on the other side, the frog, whose belly generally evokes the feminine, is penetrating into the orifice formed by the snake's mouth. What we have is a sexual act (not to be seen) which alludes equally to the imaginary scene in which the child voyeur is conceived and to a strange act in which the sexual roles are inverted. This copulation is described as 'birth the wrong way round'. Let us recall Virginia's words when she tells [in *Moments of Being*] how she was sexually attacked by George: the moment of her birth retreats millions of years backwards and does resemble a backwards birth. The inversions of the temporal movement or the sexual act are inscribed as attempts to efface the monstrous moment and begin the story again. So this passage refers to the rape scene which appears as a fundamental trauma in the *œuvre* of Virginia Woolf. It also unveils a fear of childbirth as great as that of the sexual act.

This cannot fail to evoke also what Victorian morality considered an 'inversion', namely homosexuality – or the fear of 'inverted' sexuality in three words: lust, perversion, coward(ice), which inscribe a significant trajectory (fear of perversion / fear of lust). Might it not be that what Virginia Woolf wanted to repress most severely in herself was her homosexuality?

To return to the text: the horror increases further when Giles wipes out the monstrous couple beneath his white-clad foot. Whiteness of virginity filthy with blood and rape; murderness and bloodolent virginity. Virginia, the violated virgin; this name, Virginia Woolf, traces a terrible destiny: the wolf virgin; this name was a bearer of violence. Why did she agree to take her husband's name, when it marked her as a virgin vowed to the wolf, or as a raped woman – is it a question of masochism, or of an unconscious fascination for a traumatic scene as ineffaceable as a name?

It cannot be swallowed. Everything crystallises around the impossibility of swallowing which belongs to the snake; the parallel is constructed from two words, swallow/die – as if, yoked to the scene of rape, there reappeared the fantasy of the death of the mother (frog). What is more, 'swallow' has to be linked to a motif which covers the text all over in the form of swallows; we could call this the displaced and sublimated expression of the traumatic scene across a poetic leitmotif.

In terms of biography, we cannot but think of Virginia's anorexic symptoms when she was sliding once more into madness. It does look as though these symptoms can be thought to relate as much to death as to rape. The act of absorbing food seemed repulsive to her – but what is the link between swallowing through the mouth and swallowing through the sexual organ? And how should we explain the fact that in her writing it is often the eyes which absorb in place of mouths?

If we regroup all the elements of this scene, we observe that it is organised into two phases: each stage, which is connoted as sexual, is orchestrated and unveiled as if behind any copulation lay the outline of a monstrous coupling. . . .

This is the missing fragment of the puzzle; it was written in her last book just before her death, at the end of her road, and as though at the extreme point of an *œuvre* whose concerns it decisively reveals. I don't think there is any passage more revealing of the Woolfian rips, but this does not imply that it is completely isolated, for there are hundreds of bits transcribed fleetingly, in two or three words, in many novels. The question is linked to crossing the threshold of the mouth or any other orifice – it might, for instance, be the threshold of a door or a window. . . .

Why does Mrs Ramsay tell us that the windows must be open and the doors closed? What threshold is thereby blocked in advance? In *The Voyage Out*, Rachel dies on the threshold of life, on the threshold of sexual flowering: what are we then to think of the title, which, as I suggested, signals a push towards the exterior, an impulse towards the outside? Might there not be a link between this crossing and Septimus's gesture, in *Mrs Dalloway*, when he throws himself out of the window – not to forget Virginia's own gesture when she tried to kill herself by throwing herself out of the window? These suicides are like a violent, fatal crossing of a threshold which blocks the space of life, like a voyage out; this trajectory enables us to understand the connection between the impossibility of getting over a threshold and the wish to die.

Let us look at the question of the threshold from different angles. It can have to do with the threshold in the sexual act – in other words, of a moment of tipping over – or, in the area of literature, it can have to do with the culminating point of a text which would issue in the dénouement of a plot which does not exist – since, as Woolf says in *Between the Acts*, the plot does not matter.

But there is no dénouement, or disentanglement, in the strict sense, because nothing has been tangled up. For example, the death of the mother in *To the Lighthouse* does not represent a dramatic moment since it is not the nodal point of the action: it occurs between parentheses, as though it ought to go unnoticed.

From threshold to threshold, the continuity goes on, with no spectacular events, for the substance of the novels is not made of 'action'; rather, the text unfolds in the intervals. It is composed of a series of acts of everyday life – or rather of what is lived in the interiority of each character, in the flux of their thoughts, between their acts. . . .

But in this moment of unswallowable tension which I have tried to analyse, something else happens: 'But it was action. Action relieved him.' This is *passage à l'action* in its most active and brutal mode; it provokes a dénouement by putting an end to the monstrous trance in the most realistic way possible: a foot wipes out the two animals. But this is not 'the' dénouement, since the text continues according to the same rhythm. However, the book's title refers, I think, to those moments of horror which take place between the acts, and quite precisely at the heart of the book, and it is this that marks them out as textual trauma at the centre of the work. 'Action' is between two, in the interval, between the parentheses, just where we do not expect it in a theatrical performance – it rises up from the depths of the unconscious cavern [*antre*, homonymic with *entre*, 'between'] where it was hiding. This is why it first appears as an invisible scene, not to be seen: but it is also the repressed scene which surges up again just when you weren't expecting it. The bracketing off also designates the healthy repression which enables the psyche not to die; the monstrous act(ion) is censured and becomes an act; that leads to the sublimation which is that of Acts performed on a theatrical stage.

The intolerable pain must be mitigated; and just as 'mother' becomes 'moth', so 'action' becomes 'act': thus the two most dramatic moments of Virginia Woolf's life, namely her mother's death and the sexual aggressions she suffered, are symbolised, reduced, displaced and played out in the space of literature.

But we could ask now in what other ways the text protects itself from the intolerable pain caused by violent scenes. . . .

Cut-out fragments

One solution is to flee by prolonging sleep. Orlando suffers a mortal blow when his beloved Sasha, after having betrayed him (or her), runs away and disappears forever. Orlando then sinks into a sleep which lasts for days and days and resembles a foetal stasis, from the time when the beloved mother was wholly present.

This absolves Orlando from eating and drinking, and it is reminiscent of the state in which Virginia was when she had her breakdown and became once more a tiny baby, in bed, incapable of feeding herself.

There is a violent counterpoint to sleep in the ubiquitous presence of the knife, which appears in practically all the novels. In *Mrs Dalloway*, for instance, Peter Walsh plays with the blade of his penknife all the time. Women, on the other hand, more often have scissors; this is the case with Rezia, Septimus's wife, or even with Mrs Ramsay, whose husband is compared to a blade or knife (TL, p. 10). Their son James, moreover, holds a knife throughout the text and his fantasy is to use it to strike his father to the heart, which amounts to accomplishing his (Oedipal, symbolic) murder.

This blade is a constant threat; it is related to the death drive. But its violence does not strike directly; it is displaced; it is for this reason that Peter Walsh's knife is not put into action, but Septimus dies all the same, not killed by the knife which he sees a few seconds before his death, but impaled horribly on the spikes of the railing below the window from which he jumped. On the other hand, while James Ramsay does not accomplish his secret desire with regard to the father, still death does not spare Mrs Ramsay. It seems that the one for whom violence is (in imagination) destined is not the one who suffers it. The violence addressed to the other is often turned back against the self and so spares the culprit (what culprit?); it leads to suicide.

The subterranean pressure is intense: knives and scissors increase the threat of death and rape which, over and above the moments in the text when it is exposed, seems to be continually hanging there. However, the repressed and invisible violence concentrated at the heart of the text suddenly causes the mirror to shatter in silence. The explosion sends splinters everywhere.

Fragments which are sometimes extremely sharp are thrown out in all directions, as with the hook seized from the gaping mouth of a fish when Mr Ramsay is on the way to the lighthouse with his children. The event looks trivial, but it is related to an almost identical fragment to be found in *Between the Acts*, where the monstrous scene sends out its echoes into other pages. It begins with a story in *The Times*, a rape: 'Then one of the troopers removed part of her clothing, and she screamed and hit him about the face' (BA, p. 19): so here is another version of what is so familiar to us.

The mirror breaks into thousands of scattered pieces: 'And the mirror – that I lent her. My mother's. Cracked' (BA, p. 133). Raped body and death of the mother, searing grief, which cause the self to crumble and the text to burst apart. The absence of the maternal figure, or the memory of a sexual attack, forbid the text from being organised otherwise than around the void.

The broken mirrors multiply. In *Orlando*, the queen has no sooner grasped in her hand a mirror reflecting a couple of amorous adolescents when she smashes it on the ground in anger. The theme of the broken mirror is related to that of the Great Frost: so that when Sasha betrays Orlando, the ice whose opaqueness was a bad omen cracks all over: the thaw had 'burst the ice asunder with such vehemence that it swept the huge and massy fragments furiously apart' (O, p. 39). At the same moment, the oak trees are split and torn apart. As we can see, the simultaneity of these events is significant: the oak tree collapses and the mirror shatters to bits when the centre is attacked, in other words when the loved woman disappears.

Fragments of the mirror, and the body in pieces. In Woolf's novels, bodies are not described in the plenitude of beauty – except for the face of Mrs Ramsay, for example. Often, the head seems to have no body, it is cut off from the ensemble. The representation of a physique becomes almost expressionist when in *Between the Acts*, 'a gigantic ear attached to a gigantic head' (BA, p. 133) makes its appearance. To understand the phenomenon better, we need to refer to the passage in *Moments of Being* when the image Virginia is looking at in the mirror is doubled into a monstrous animal. Her body is at once 'bird of paradise' and 'vulture'. The spiritual animal, so to speak, expressed in the luminous beauty of a face, is cut off from sensual love – called lust.

What then of the question of sexuality as it appears in the writings? It has too often and too swiftly been decreed that Virginia Woolf was 'frigid'. This is wrong – we need only read a few pages of *Moments of Being* to be convinced of this. However, if we draw together all the elements I have analysed, we can see that the double trauma has left some traces. Sensuality and *jouissance* do not necessarily mean genitality. Sexual horror is confined to a number of highly specific spaces.

And indeed, whenever the text comes close to something which might resemble a penis, the revulsion is extreme, and I think this is how we should read the passage in which Orlando is looking at Joe's finger without a nail: 'In its place was a pink roll of flesh. It was so repulsive that Orlando almost fainted' (O. p. 38).

The body, which seems to be on the far side of the mirror stage, is decomposed into disparate pieces: 'Cut off from their bodies, their eyes smiled, their bodiless eyes, at their eyes in the glass (BA, p. 56). This sentence gives the impression of the multiplication of broken bits of eyes in the shattered mirror. 'I' and 'eye' being homonyms, the eyes also refer to the self when it crumbles apart into pieces in one of the concluding passages of *Orlando*, where the heroine watches 'a great variety of selves'

73

present themselves in order, all different aspects of her personality (O, p. 193).

This is the direction in which we should look to decipher the structure of *The Waves*: the characters, who cannot be differentiated from one another because they are not individuated, are part of the same whole, as though they were just the six different facets of just one personality – or even of one single, united family (Leslie, Julia and their many children): a happy family broken into thousands of pieces after the mother's death. Woolf's *œuvre* reveals the search for a unity of the self as well as that of the happy family – relegated to the Eden of childhood.

Her writing reposes, if that is the word, on the principle of the broken mirror, for the novels are constructed like collages made from 'bits and pieces': this is a vast bric-à-brac in which chunks of letters, words, sentences are juxtaposed in a crude heaping, as if the higgledy-piggledy side of the real were reproduced in the way that it seems to offer itself to us. For instance, the audience's comments at the end of the play of *Between the Acts* are delivered to us all mixed up: 'I thought it brilliantly clever. . . . O my dear, I thought it utter bosh. Did *you* understand the meaning? Well, he said she meant we all act all parts. . . . He said, too, if I caught his meaning, Nature takes part. . . . Then there was the idiot . . .' (BA, p. 143).

This page is like a montage, and its technique could be related to that of film. The construction of *The Waves* shares the same principle: the pieces of poetry intercalated between the interior monologues are assembled in a collage made out of different kinds of rhythm. Similar observations can be made about *Between the Acts*: vignettes from the history of England are juxtaposed and they merge with whatever comes up, refrains, noises, bits of conversation (of which the quotation above is an example), dips into individual characters, songs, etc.

If we continue the catalogue and go on to *Jacob's Room*, there are the same characteristics: the novel uses the innumerable sparkling facets of the life of a young man. In *The Years*, fragments of the Pargiter family's life are glimpsed amid the flux of History, but they are never reflected whole. And if Woolf's technique consists in using light, it also rests on a fragmentation of the approach which grasps each character in the multiplicity of their aspects because, so this writer thought, it is difficult to know others other than as beings who are fragmented and discontinuous.

But we could ask whether the writer is really making collages – or cut-outs? The first few pages of *To the Lighthouse* are highly significant: James Ramsay is occupied in cutting out the pictures from a catalogue. The scene is in no way anodyne, but relates to one of the themes I have already invoked: the knife–scissor. This is used first to cut out the pictures: it is only at a second stage that these will be juxtaposed as in a mosaic: it is necessary to 'cut people up . . . and stick them up again' (TL, p. 9).

These lines will enable us to interrogate the author's way of writing: what then is the use of the scissors and knives? Does it have to do with an imaginary fragmentation and isolation of sensations by means of the scissors? Does she cut up life as it is lived so as to bring together the bits, giving them back a unity which only writing can create, as she says in *Moments of Being*, or does she live like Mrs Dalloway, who 'cut like a knife through everything' (MD, p. 11)?

Why is the texture of the novels chopped up so finely by an excess of punctuation – isn't this in a way the reverse of the wave-rhythm, which is so seductive when you first read this writer? Yet her use of punctuation cuts both ways, so to speak, for she cuts the page into an infinite number of fragments. The words are surrounded, aureoled and separated from each other by a vast quantity of ellipses, exclamation marks, question marks, as in some parts of *Mrs Dalloway*. This also produces an impression of withholding, like poetic inspiration which does not want to yield to its first impulse and is taking the measure of each of its steps. The overflow is contained, only a 'jet of water' (MD, p. 17) reaches, there is a cut.

But it is really in the rhythm that the key to the writing and the cutting up is to be found, for even if the centre has been shown to be 'empty, empty' (MD, p. 117), that does not mean that the text goes off in all directions, out of control. Woolf is a brilliant writer, and her texts cannot be 'normalised' through a reductive reading. The unity and originality of each one depends on a rhythm through which the structure, completely mobile and supple, can be outlined. So the knife is also used to cut up the structure of the novels.

The cutting machines are numerous and take many forms. They are active, and it is out of the rhythm of a machine going 'chuff, chuff, chuff' regularly in the text that everything in *Between the Acts* is cemented into a coherent whole. The sound is as relentless as that of an insect, as regular as that of a combine harvester. The machine evokes all machines, and it functions via a three-beat rhythm, Virginia Woolf's preferred measure (the three rays of the lighthouse, etc.). This creates a melody in the very texture of the novel. We also think of the machine of classical theatre, or the writing machine [*machine à écrire*, a typewriter] – in other words, the machine for text production.

This is how the text is fabricated, from cuts, intervals, or interludes which literally enable us to take a breath (chuff, chuff, chuff), and to get going again, like a well-running machine, as regularly as a clock.

The rhythm then becomes 'tick, tick, tick', like the beating which animates the body: this is how it is described in *Mrs Dalloway*, in which the novel's structure also rests on a regular rhythm, that of the clock. There is no chapter division, but the clock-sectioning outlines a framework: the clocks sound at regular intervals, as a theme structures a musical work: 'Shredding and slicing, dividing and subdividing, the clocks of Harley Street nibbled at the June day' (MD, p. 113).

This novel which takes place throughout a beautiful day in June only ends when the party is over at Mrs Dalloway's house of light; and we hear Big Ben ringing out 'first a musical prelude; then the irrevocable hour' (MD, p. 128), not systematically every hour, but according to how the characters are living the moment of the day. For example, the hours and half-hours of the morning are clearly punctuated by Big Ben because this corresponds to precise activities. But at night, during the party, time runs on unseen, very fast.

The division of *The Waves*, which has elicited much admiring commentary, does not seem successful to me, although from a certain point of view it could be hailed as a technical feat. The book's title indicates its structure in a wholly readable way. But that is just the problem: the procedure is too systematic, with the text artificially divided into waves, as though by some machine that had lost the sense of rhythm and no longer knew how to breathe. It resembles a décor made of pasteboard waves activated by a cold mechanism and pretending to go up and down. The title makes do with miming what is not there in the text: the supple, living rhythm of the wave which animates other novels does not manage to spread itself across the surface of this text.

When the novels are successful, that is due to the fact that the rhythm they rest on is double: on the one hand, there is for instance the sharp rhythm of Big Ben (in *Mrs Dalloway*), or 'tick, tick, tick' (in *Between the Acts*), and on the other, there are fragments of sentences relentlessly repeated, underneath, as gently as the murmur of ebb and flow.

In *Between the Acts*, the swallow motif has the refrain 'Swallow, my sister, O sister swallow' (BA, p. 87); or again, each time Big Ben sounds, we hear the echo, very softly: 'The leaden circles dissolve in the air'. It is as though these repetitions were buried between the lines, not immediately perceptible, or rather, addressing the reader from the depths as though they were the intimate vibration of the text, the counterpoint of the clock. The utter originality of Virginia Woolf's novelistic *œuvre* is born of a meeting between an external division according to a rhythm and an entirely internal rhythm: but could we say that this means that it offers itself as women's writing [*écriture féminine*] – and what would that mean?

Note

References within the text abbreviate titles to their initials. Granada editions have been used throughout, except for *Mrs Dalloway*, where references are to the 1975 Penguin. Translator's notes are included in square brackets.

5 Narrative Structure(s) and Female Development: the Case of *Mrs Dalloway**

ELIZABETH ABEL

This piece reads *Mrs Dalloway* for its implied commentary on a standard story of the development of subjectivity, and suggests that Woolf is able in the 1920s to let in more of the feminist criticisms of a dominantly masculine line, in the form of her subversively planted alternative plotting of the less obvious female stories, than could previously have found a place. Juxtaposing Woolf and Freud, two story-tellers in different modes, Abel suggests that narrative and gender are indissolubly related: no story is neutral. She is interested in the way that *Mrs Dalloway*, with its intimations of passionate female attachments in early life, may be hinting at 'a larger female story of natural existence abruptly curtailed' and valorising 'a spontaneous homosexual love over the inhibitions of imposed heterosexuality'.

I wish you were a Kangaroo and had a pouch for small Kangaroos to creep to.

(Virginia Stephen to Violet Dickinson, 4 June (?), 1903)

Our insight into this early, pre-Oedipus, phase comes to us as a surprise, like the discovery, in another field, of the Minoan–Mycenean civilization behind the civilization of Greece.

(Sigmund Freud, 'Female Sexuality', 1931)

We all know Virginia Woolf disliked the fixity of plot: 'This appalling narrative business of the realist', she called it.[1] Yet like all writers of fiction, she inevitably invoked narrative patterns in her work, if only to disrupt them or reveal their insignificance. In *Mrs Dalloway*, a transitional work between the straightforward narrative of an early novel like *The Voyage Out* and the experimental structure of a late work like *The Waves*,

---

* Reprinted from Elizabeth Abel, Marianne Hirsch and Elizabeth Langland (eds), *The Voyage In: Fictions of Female Development* (Hanover: University of New England Press, 1983), pp. 161–85.

Woolf superimposes the outlines of multiple, familiar yet altered plots that dispel the constraints of a unitary plan, diffuse the chronological framework of the single day in June, and enable an iconoclastic plot to weave its course covertly through the narrative grid. In this palimpsestic layering of plots, *Mrs Dalloway* conforms to Gilbert and Gubar's characterization of the typically female text as one which both inscribes and hides its subversive impulses.[2]

The story of female development in *Mrs Dalloway*, a novel planned such that 'every scene would build up the idea of C[larissa]'s character',[3] is a clandestine story that remains almost untold, that resists direct narration and coherent narrative shape. Both intrinsically disjointed and textually dispersed and disguised, it is the novel's buried story. The fractured developmental plot reflects the encounter of gender with narrative form and adumbrates the psychoanalytic story of female development, a story Freud and Woolf devised concurrently and separately, and published simultaneously in 1925. The structure of Woolf's developmental story and its status in the novel illustrate distinctive features of female experience and female plots.

Woolf repeatedly acknowledged differences between male and female writing, detecting the influence of gender in fictional voice and plot. While insisting that the creative mind must be androgynous, incandescent, and unimpeded by personal grievance, she nevertheless affirmed that differences between male and female experience would naturally emerge in distinctive fictional shapes. She claims,

> No one will admit that he can possibly mistake a novel written by a man for a novel written by a woman. There is the obvious and enormous difference of experience in the first place . . . And finally . . . there rises for consideration the very difficult question of the difference between the man's and the woman's view of what constitutes the importance of any subject. From this spring not only marked differences of plot and incident, but infinite differences in selection, method and style.[4]

The experience that shapes the female plot skews the woman novelist's relationship to narrative tradition; this oblique relationship may further mold the female text. In a remarkable passage in *A Room of One's Own*, Woolf describes one way in which the difference in experience can affect the logic of the female text:

> And since a novel has this correspondence to real life, its values are to some extent those of real life. But it is obvious that the values of women differ very often from the values which have been made by the other sex; naturally, this is so. Yet it is the masculine values that

prevail . . . And these values are inevitably transferred from life to fiction. This is an important book, the critic assumes, because it deals with war. This is an insignificant book because it deals with the feelings of women in a drawing-room. A scene in a battlefield is more important than a scene in a shop – everywhere and much more subtly the difference of value persists. The whole structure, therefore, of the early nineteenth-century novel was raised, if one was a woman, by a mind which was slightly pulled from the straight, and made to alter its clear vision in deference to external authority. . . . The writer was meeting criticism . . . She met that criticism as her temperament dictated, with docility and diffidence, or with anger and emphasis. It does not matter which it was; she was thinking of something other than the thing itself. . . . She had altered her values in deference to the opinions of others.[5]

Woolf explicitly parallels the dominance of male over female values in literature and life, while implying a different hierarchy that further complicates the woman novelist's task. By contrasting the 'values of women' with those which 'have been made by the other sex', Woolf suggests the primacy of female values as products of nature rather than culture, and of the named sex rather than the 'other' one. No longer the conventionally 'second' sex, women here appear the source of intrinsic and primary values. In the realm of culture, however, masculine values prevail and deflect the vision of the woman novelist, in setting a duality into the female narrative, turned Janus-like toward the responses of both self and other. This schizoid perspective can fracture the female text. The space between emphasis and undertone, a space that is apparent in Woolf's own text, may also be manifested in the gap between a plot that is shaped to confirm expectations and a subplot at odds with this accommodation. If the early nineteenth-century woman novelist betrayed her discomfort with male evaluation by overt protestation or compliance, the early twentieth-century woman novelist, more aware of this dilemma, may encode as a subtext the stories she wishes yet fears to tell.

Feminist literary criticism, Elaine Showalter states, presents us with 'a radical alteration of our vision, a demand that we see meaning in what has previously been empty space. The orthodox plot recedes, and another plot, hitherto submerged in the anonymity of the background, stands out in bold relief like a thumbprint.'[6] The excavation of buried plots in women's texts has revealed an enduring, if recessive, narrative concern with the story of mothers and daughters – with the 'lost tradition', as the title of one anthology names it, or, in psychoanalytic terminology, with the 'pre-Oedipal' relationship, the early symbiotic female bond that both predates and coexists with the heterosexual orientation toward the father and his substitutes. Frequently, the subtleties of mother–daughter alignments, for which few narrative conventions have been formulated, are relegated to the background of a dominant romantic or courtship plot. As women novelists increasingly exhaust or dismiss the possibilities of the romantic plot, however, they have tended to inscribe the maternal subplot more emphatically. In contemporary women's fiction, this subplot is often dominant; but in the fiction of the 1920s, a particularly fruitful decade for women and women's writing, the plot of female bonding began to vie repeatedly with the plot of heterosexual love. Woolf, Colette and Cather highlighted aspects of the mother–daughter narrative in works such as *My Mother's House* (1922), *To the Lighthouse* (1927), *Break of Day* (1928), *Sido* (1929) and 'Old Mrs Harris' (1932).[7] In *Mrs Dalloway*, written two years before *To the Lighthouse*, Woolf structures her heroine's development, the recessive narrative of her novel, as a story of pre-Oedipal attachment and loss.

In his essay 'Female Sexuality', Freud parallels the pre-Oedipal phase of female development to the allegedly matriarchal civilization lying behind that of classical Greece, presumably associated here with its most famous drama; his analogy offers a trope for the psychological and textual strata of *Mrs Dalloway*.[8] For Freud conflates, through the spatial and temporal meanings of the word 'behind' (*hinter*), notions of evolution with those of static position. Clarissa Dalloway's recollected development proceeds from an emotionally pre-Oedipal female-centered natural world to the heterosexual male-dominated social world, a movement, Woolf implies, that recapitulates the broader sweep of history from matriarchal to patriarchal orientation. But the textual locus of this development, to revert to the archaeological implications of Freud's image, is a buried *sub*text that endures throughout the domestic and romantic plots in the foreground: the metaphors of palimpsest and cultural strata coincide here. The interconnections of female development, historical progress, and narrative structure are captured in Freud's image of a pre-Oedipal world underlying the individual and cultural origins we conventionally assign the names Oedipus and Athens.

Woolf embeds her radical developmental plot in a narrative matrix pervaded by gentler acts of revision; defining the place of this recessive plot requires some awareness of the larger structure. The narrative present, patterned as the sequence of a day, both recalls the structure of *Ulysses*, which Woolf completed reading as she began *Mrs Dalloway*, and offers a female counterpart to Joyce's adaptation of an epic form.[9] *Mrs Dalloway* inverts the hierarchy Woolf laments in *A Room of One's Own*. Her foregrounded domestic plot unfolds precisely in shops and drawing rooms rather than on battlefields, and substitutes for epic quest and conquest the traditionally feminine project of giving a party, of constructing social harmony through affiliation rather than conflict; the potentially epic plot of the soldier returned from war is demoted to the tragic subplot centering on Septimus Warren Smith. By echoing the structure of *Ulysses* in the narrative foreground of her text, Woolf revises a revision of the epic to accommodate the values and experience of women while cloaking the more subversive priorities explored in the covert developmental tale.

A romantic plot, which provides the dominant structure for the past in *Mrs Dalloway*, also obscures the story of Clarissa's development. Here again, Woolf revises a traditional narrative pattern, the courtship plot perfected by Woolf's elected 'foremother', Jane Austen. Woolf simultaneously invokes and dismisses Austen's narrative model through Clarissa's mistaken impression that her future husband is named Wickham. This slight, if self-conscious, clue to a precursor assumes greater import in the light of Woolf's lifelong admiration for Austen and Woolf's efforts to reconstruct this 'most perfect artist among women' in her literary daughter's image; these efforts structure Woolf's essay on Austen, written shortly after *Mrs Dalloway*.[10] Woolf's treatment of the romantic plot in *Mrs Dalloway* reveals the temporal boundaries of Austen's narratives, which cover primarily the courtship period and inevitably culminate in happy marriages. Woolf condenses the expanded moment that constitutes an Austen novel and locates it in a remembered scene thirty years prior to the present of her narrative, decentering and unraveling Austen's plot. Marriage in *Mrs Dalloway* provides impetus rather than closure to the courtship plot, dissolved into a retrospective oscillation between two alluring possibilities as Clarissa continues to replay the choice she made thirty years before. The courtship plot in this novel is both evoked through memories of the past and indefinitely suspended in the present, completed when the narrative begins and incomplete when the narrative ends, sustained as a narrative thread by Clarissa's enduring uncertainty. The novel provides no resolution to this internalized version of the plot; the final scene presents Clarissa through Peter Walsh's amorous eyes and allies Richard Dalloway with his daughter. The elongated courtship plot, the imperfectly resolved emotional triangle, becomes a screen for the developmental story that unfolds in fragments of memory, unexplained interstices between events, and narrative asides and interludes.

When Woolf discovered how to enrich her characterization by digging 'beautiful caves' into her characters' pasts,[11] her own geological image for the temporal strata of *Mrs Dalloway*, she chose with precision the consciousness through which to reveal specific segments of the past. Although Clarissa vacillates emotionally between the allure of Peter and that of Richard, she remembers Peter's courtship only glancingly; the burden of that plot is carried by Peter, through whose memories Woolf relates the slow and tortured end of the relation with Clarissa. Clarissa's memories, by contrast, focus more exclusively on the general ambience of Bourton, her childhood home, and her love for Sally Seton. Significantly absent from these memories is Richard Dalloway, whose courtship of Clarissa is presented exclusively through Peter's painful recollections. Clarissa thinks of Richard only in the present, not at the peak of a romantic relationship. Through this narrative distribution, Woolf constructs two diversified poles structuring the flux of Clarissa's consciousness. Bourton is to Clarissa a pastoral female world spatially and temporally disjunct from marriage and the sociopolitical world of (Richard's) London. The fluid passage of consciousness between these poles conceals a radical schism.

Though the Bourton scenes Clarissa remembers span a period of several years, they are absorbed by a single emotional climate that creates a constant backdrop to the foregrounded day in June. Woolf excises all narrative connections between these contrasting extended moments. She provides no account of intervening events; Clarissa's marriage, childbirth, the move and adjustment to London. And she indicates the disjunction in Clarissa's experience by noting that the London hostess never returns to Bourton, which now significantly belongs to a male relative, and almost never sees Sally Seton, now the unfamiliar Lady Rosseter. Clarissa's life in London is devoid of intimate female bonds: she is excluded from lunch at Lady Bruton's and she vies with Miss Kilman for her own daughter's allegiance. Woolf structures Clarissa's development as a stark binary opposition between past and present, nature and culture, feminine and masculine dispensations – the split implicit in Woolf's later claim that 'the values of women differ very often from the values which have been made by the other sex'. Versions of this opposition reverberate throughout the novel in rhetorical and narrative juxtapositions. The developmental plot, which slides beneath the more familiar romantic plot through the gap between Peter's and Clarissa's memories, exists as two contrasting moments and the silence adjoining and dividing them.

Woolf endows these moments with symbolic resonance by a meticulous strategy of narrative exclusions that juxtaposes eras split by thirty years and omits Clarissa's childhood from the novel's temporal frame. There is no past in *Mrs Dalloway* anterior to Clarissa's adolescence at Bourton. Within this selective scheme, the earliest remembered scenes become homologous to a conventional narrative point of departure: the description of formative childhood years. The emotional tenor of these scenes, moreover, suggests their representation of deferred childhood desire. Clarissa's earliest narrated memories focus on Sally's arrival at Bourton, an arrival that infuses the formal, repressive atmosphere with a vibrant female energy. The only picture of Clarissa's early childhood sketched in the novel suggests a tableau of female loss: a dead mother, a dead sister, a distant father, and a stern maiden aunt, the father's sister, whose hobby of pressing flowers beneath Littré's dictionary suggests to Peter Walsh the social oppression of women, an emblem of nature ossified by language/culture. In this barren atmosphere, Sally's uninhibited warmth and sensuality immediately spark love in the eighteen-year-old Clarissa.[12] Sally replaces Clarissa's dead mother and sister, her name even echoing the sister's name, Sylvia. She nurtures Clarissa's passions and intellect, inspiring a love equal to Othello's in intensity and equivalent in absoluteness to a daughter's earliest bond with her mother, a bond too early ruptured for Clarissa as for Woolf, a bond which Woolf herself perpetually sought to recreate through intimate attachment to mother surrogates, such as Violet Dickinson: 'I wish you were a Kangaroo and had a pouch for small Kangaroos to creep to.'[13] For Clarissa, kissing Sally creates the most exquisite moment of her life, a moment of unparalleled radiance and intensity:

> The whole world might have turned upside down! The others disappeared; there she was alone with Sally. And she felt she had been given a present, wrapped up, and told just to keep it, not to look at it – a diamond, something infinitely precious, wrapped up, which, as they walked (up and down, up and down), she uncovered, or the radiance burnt through, the revelation, the religious feeling! – when old Joseph and Peter faced them.[14]

This kind of passionate attachment between women, orthodox psychoanalysts and feminists uncharacteristically agree, recaptures some aspect of the fractured mother–daughter bond.[15] Within the sequence established by the novel, this adolescent love assumes the power of the early female bond excluded from the narrative.

The moment Woolf selects to represent Clarissa's past carries the full weight of the pre-Oedipal experience that Freud discovered with such a shock substantially predates and shapes the female version of the Oedipus complex, the traumatic turn from mother to father. As French psychoanalytic theory has clarified, the Oedipus complex is less a biologically ordained event than a symbolic moment of acculturation, the moment, in Freud's words, that 'may be regarded as a victory of the race over the individual', that 'initiates all the processes that are designed to make the individual find a place in the cultural community'.[16] For both women and men, this socialization exacts renunciation, but for women this is a process of poorly compensated loss, for the boy's rewards for renouncing his mother will include a woman like the mother and full paternal privileges, while the girl's renunciation of her mother will at best be requited with a future child, but no renewed access to the lost maternal body, the first love object for girls as well as boys, and no acquisition of paternal power. In *Mrs Dalloway*, Woolf encapsulates an image of the brusque and painful turn that, whenever it occurs, abruptly terminates the earliest stage of female development and defines the moment of acculturation as a moment of obstruction.

Woolf organizes the developmental plot such that Clarissa's love for Sally precedes her allegiances to men; the two women 'spoke of marriage always as a catastrophe' (p. 50). Clarissa perceives Peter in this period primarily as an irritating intruder. The scene that Clarissa most vividly remembers, the scene of Sally Seton's kiss, is rudely interrupted by Peter's appearance.[17] Both the action and the language of this scene hint at psychological allegory. The moment of exclusive female connection is shattered by masculine intervention, a rupture signaled typographically by Woolf's characteristic dash. Clarissa's response to this intrusion images an absolute and arbitrary termination: 'It was like running one's face against a granite wall in the darkness! It was shocking; it was horrible!' (p. 53). Clarissa's perception of Peter's motives – 'she felt his hostility; his jealousy; his determination to break into their comradeship' – suggests an Oedipal configuration: the jealous male attempting to rupture the exclusive female bond, insisting on the transference of attachment to the man, demanding hererosexuality. For women this configuration institutes a break as decisive and unyielding as a granite wall. Clarissa's revenge is to refuse to marry Peter and to select instead the less demanding Richard Dalloway in order to guard a portion of her psyche for the memory of Sally. Woolf herself exacts poetic justice by subjecting Peter Walsh to a transposed, inverted replay of this crucial scene when Elizabeth, thirty years later, interrupts his emotional reunion with her mother by unexpectedly opening a door (in the granite wall?), asserting by her presence the primacy of female bonds. 'Here is my Elizabeth', (p. 71) Clarissa announces to the disconcerted Peter, the possessive pronoun he finds so extraneous accentuating the intimacy of the mother–daughter tie.

Clarissa resists the wrenching, requisite shift from pre-Oedipal to Oedipal orientation, yet she submits in practice if not totally in feeling. The extent of the disjunction she undergoes is only apparent in the bifurcated settings of her history, the images reiterating radical divides, the gaps slyly inserted in the narrative. The most striking of these images and gaps concern Clarissa's sister Sylvia, a shadowy and seemingly gratuitous character, apparently created just to be destroyed. Her death, her only action in the novel, is recalled by Peter rather than by Clarissa and is related in two sentences. This offhand presentation both implants and conceals an exaggerated echo of Clarissa's split experience. A young woman 'on the verge of life' (p. 118), Sylvia is abruptly killed by a falling tree that dramatically imposes a barrier to life in a gesture of destruction mysteriously associated with her father: '(all Justin Parry's fault – all his carelessness)' (pp. 118–19). The shocking attribution of blame is only ostensibly discounted by parentheses: recall Woolf's parenthetical accounts of human tragedy in the 'Time Passes' section of *To the Lighthouse*. The deliberate decision to indict the father contrasts with the earlier story, 'Mrs Dalloway in Bond Street', where Sylvia's death is depicted as a tranquil, vague event absorbed by nature's cyclical benevolence: 'It used, thought Clarissa, to be so simple. . . . When Sylvia died, hundreds of years ago, the yew hedges looked so lovely with the diamond webs in the mist before early church.'[18] The violence of Sylvia's death in the novel and the very incongruity between the magnitude of the charge against her father and its parenthetical presentation suggest a story intentionally withheld, forcibly deprived of its legitimate proportions, deliberately excised from the narrative yet provocatively implied in it, written both into and out of the text. This self-consciously inscribed narrative gap echoes the gap in Clarissa's own narrative, as the dramatic severance of Sylvia's life at the moment of maturity echoes the split in her sister's development. The pastoral resonance of Sylvia's name also implies a larger female story of natural existence abruptly curtailed.[19] A related narrative exclusion suggests a crucial untold tale about Clarissa's relation to her mother, remarkably unremembered and unmentioned in the novel except by a casual party guest whose brief comparison of Clarissa to her mother brings sudden tears to Clarissa's eyes. The story of the pain entailed in this loss is signaled by but placed outside the narrative in the double gesture of inclusion and exclusion that structures Woolf's narration of her heroine's development. By locating the clues to this discontinuous narrative in the marginal moments of her text, Woolf creates an inconspicuous subtext perceptible only to an altered vision.

Woolf's discrete suggestion of an intermittent plot is politically astute and aesthetically adept. Her insight into the trauma of female development does subvert the notion of organic, even growth culminating for women in marriage and motherhood, and she prudently conceals her implications of a violent adaptation. The narrative gaps also challenge the conventions of linear plot and suggest its distorted regimentation of experience, particularly the subjective experience of women. These gaps, moreover, are mimetically precise: juxtapositions represent sudden shifts, silence indicates absence and loss. Perhaps Woolf's most striking achievement, however, is her intuition of the 'plot' Freud detected in female development. Despite Woolf's obvious familiarity with the popularized aspects of Freudian theory, and despite the close association of the Hogarth Press with the Freudian *œuvre*, there can be no question of influence here, for Freud first expounded his view of a distinctively female development the year of *Mrs Dalloway's* publication.[20] Rather than influence, *Mrs Dalloway* demonstrates the common literary prefiguration of psychoanalytic doctrine, which can retroactively articulate patterns implicit in the literary text. The similarities between these fictional and psychoanalytic narratives clarify the structure of Woolf's submerged developmental plot and the power of Freud's submerged demonstration of the loss implicit in female development.

Only late in life did Freud acknowledge the fundamentally different courses of male and female development. Prior to the 1925 essay entitled 'Some Psychical Consequences of the Anatomical Distinction Between the Sexes', Freud clung, though with increasing reservations, to a view of sexual symmetry in which male and female versions of the Oedipal experience were fundamentally parallel. His growing appreciation of the pre-Oedipal stage in girls, however, finally toppled his view of parallel male and female tracks, inspiring a new formulation of the distinctively female developmental tasks. Female identity is acquired, according to this new theory, by a series of costly repressions from which the male child is exempt. The girl's developmental path is more arduous and bumpy than the boy's smoother linear route. For though the male child must repress his erotic attachment to his mother, he must undergo no change in orientation, since the mother will eventually be replaced by other women with whom he will achieve the status of the father; he suffers an arrest rather than a dislocation. The girl, in contrast, must reverse her course. Like the boy, she begins life erotically bonded with her mother in the symbiotic pre-Oedipal stage, but unlike him she must replace this orientation with a heterosexual attraction to her father. She must change the nature of her desire before renouncing it.

How, Freud repeatedly asks, does the girl accomplish this monumental shift from mother to father? Though the answers he proposes may be dubious, the persistent question indicates the magnitude of the event. The girl's entire sexuality is defined in this transition. She switches not only the object of her erotic interest, but also her erotic zone and mode, relinquishing the active, 'masculine', clitoridal sexuality focused on her mother for the passive, receptive, 'feminine', vaginal sexuality focused on her father. Freud goes so far as to call this change a 'change in her own sex', for prior to this crucial shift, 'the little girl is a little man'.[21] This comprehensive change in sexual object, organ and attitude, the shift from pre-Oedipal to Oedipal orientation, inserts a profound discontinuity into female development, which contrasts with that of 'the more fortunate man [who] has only to continue at the time of his sexual maturity the activity that he has previously carried out at the period of the early efflorescence of his sexuality'.[22] The psychosexual shift that occurs in early childhood, moreover, is often reenacted in early adulthood, for marriage typically reinstates a disruption in women's experience, confined until recently to a largely female sphere prior to the heterosexual contract of marriage.[23]

The circuitous route to female identity, Freud acknowledged, is uniquely demanding and debilitating; 'a comparison with what happens with boys tells us that the development of a little girl into a normal woman is more difficult and more complicated, since it includes two extra tasks [the change of sexual object and organ]. to which there is nothing corresponding in the development of a man'.[24] No woman completes this difficult process unscathed. Freud outlines three developmental paths for women; all exact a substantial toll. If she follows the first, the girl negotiates the shift from mother to father by accepting the unwelcome 'fact' of her castration, detected in comparisons between herself and little boys. Mortified by this discovery of inferiority, aware she can no longer compete for her mother with her better endowed brother, she renounces her active sexual orientation toward her mother, deprived like herself of the valued sexual organ, and accepts a passive orientation toward the superior father. Unfortunately, the girl's renunciation of active sexuality normally entails repressing 'a good part of her sexual trends in general', and this route leads to sexual inhibition or neurosis, to 'a general revulsion from sexuality'.[25] If she chooses the second path, the girl simply refuses this renunciation, clings to her 'threatened masculinity', struggles to preserve her active orientation toward her mother, and develops what Freud calls a 'masculinity complex', which often finds expression in homosexuality.[26] Only the third 'very circuitous' path leads to the 'normal female attitude' in which the girl takes her father as the object of her passive eroticism and enters the female Oedipus complex. Curiously, however, Freud never describes this route, which turns out to be only a less damaging version of the first path toward inhibition and neurosis.[27] To the extent that her sexuality survives her 'catastrophic' repression of her 'masculine' desire for her mother, the girl will be able to complete her turn to her father and seal her femininity by desiring his baby. 'Normal' femininity is thus a fragile, tenuous proposition; no unique course is prescribed for its achievement. Freud's most optimistic prognosis assumes a doubly hypothetical, negative form: 'If too much is not lost in the course of it [development] through repression, this femininity may turn out to be normal.'[28] The achievement of this femininity, moreover, is only the first stage, for the female Oedipus complex, like the male, must itself be overcome, and the hard-won desire for the father renounced and transferred to other men. Female development thus entails a double disappointment in contrast with the single renunciation required of men. No wonder Freud concludes the last of his essay on femininity by contrasting the youthful flexibility of a thirty-year-old male with the psychical rigidity of a woman the same age:

87

Her libido has taken up final positions and seems incapable of exchanging them for others. There are no paths open to further development; it is as though the whole process had already run its course and remains thenceforward insusceptible to influence – as though, indeed, the difficult development to femininity had exhausted the possibilities of the person concerned.[29]

In *Mrs Dalloway*, Woolf suggests the developmental turn that Freud accentuates in his studies of femininity. The narratives they sketch share a radically foreshortened notion of development, condensed for Freud into a few childhood years, focused for Woolf in a single emotional shift. Both narratives eschew the developmental scope traditionally assumed by fiction and psychology, committed to detailing the unfolding of a life, and both stress the discontinuities specific to female development. Woolf, moreover, portrays the sexual and emotional calcification that Freud suggests is the toll of 'normal' development. Clarissa is explicit about her unimpassioned response to men, a response she perceives as a failure and a lack, guarding of virginity through motherhood and marriage. Her emotional and physical self-containment is represented by her narrow attic bed, where she reads Baron Marbot's memoirs of the retreat from Moscow, a victory achieved by icy withdrawal.[30] The association of her bed with a grave – 'Narrower and narrower would her bed be' (pp. 45–6) – links her adult sexuality with death. Yet, in a passage of extraordinary erotic writing, Woolf contrasts the description of the narrow bed with Clarissa's passionate responses to women, implying through this juxtaposition the cost of the pivotal development choice:

Yet she could not resist sometimes yielding to the charm of a woman, not a girl, of a woman confessing, as to her they often did, some scrape, some folly . . . she did undoubtedly then feel what men felt. Only for a moment; but it was enough. It was a sudden revelation, a tinge like a blush which one tried to check and then, as it spread, one yielded to its expansion, and rushed to the farthest verge and there quivered and felt the world come closer, swollen with some astonishing significance, some pressure of rapture, which split its thin skin and gushed and poured with an extraordinary alleviation over the cracks and sores! Then, for that moment, she had seen an illumination; a match burning in a crocus; an inner meaning almost expressed. But the close withdrew; the hard softened. It was over – the moment. Against such moments (with women too) there contrasted (as she laid her hat down) the bed and Baron Marbot and the candle half-burnt.

(pp. 46–7)

Woolf's language renders a passion that is actively directed toward women, and implicitly 'masculine' in attitude and character, yet also receptive and 'feminine', epitomized in the image of the match in the crocus, an emblem of active female desire that conflates Freud's sexual dichotomies. The power of the passage derives in part from the intermeshed male and female imagery, and the interwoven languages of sex and mysticism, a *mélange* that recurs in Clarissa's memory of Sally Seton's kiss. Fusion – of male and female, active and passive, sacred and profane – is at the heart of this erotic experience. Freud's opposition of active, 'masculine', pre-Oedipal sexuality to the passive, 'feminine', Oedipal norm denies the basis for this integration. Clarissa's momentary illumination is enabled only by the sexual orientation Freud devalues as (initially) immature and (subsequently) deviant. Woolf's passage suggests the potential completeness Freud denies the pre-Oedipal realm and calls into question the differentiation of normal from aberrant sexuality. The stark contrast between the passionate moment and the narrow bed, another juxtaposition that conceals a schism between two radically different sexual worlds, subverts the oppposition normal/abnormal. Woolf here elevates Freud's second developmental path over the costly route toward 'normal femininity', as she valorizes a spontaneous homosexual love over the inhibitions of imposed heterosexuality.

As the passage continues, the gap between the sexual options emblematized by the moment and the bed evolves into the familiar split between Sally Seton and Richard Dalloway, the split that structures the developmental plot. The allegorical image of the bed leads to a more concrete description of Clarissa's reaction to her husband's return: 'if she raised her head she could just hear the click of the handle released as gently as possible by Richard, who slipped upstairs in his socks and then, as often as not, dropped his hot-water bottle and swore! How she laughed!' (p. 47). The contrast between the passionate moment with women and the narrow marital bed becomes a leap from the sublime to the (affectionately) ridiculous. Opening with the conjunction 'But', the next paragraph signals a turn away from mundanity back to 'this question of love . . . this falling in love with women' (p. 48), inaugurating Clarissa's lengthy and lyrical reminiscence of Sally Seton. The opposition between Clarissa's relationship with men and women modulates to the split between her present and her past, her orientation and emotional capacities on both sides of the Oedipal divide. Woolf, like Freud, reveals the cost of female development, but she inscribes a far more graphic image of the loss entailed, questions its necessity, and indicates the price of equating female development with acculturation through the rites of passage established by the Oedipus complex.

These are radical claims, and Woolf suggests them indirectly. In addition to her use of juxtaposition as a narrative and rhetorical strategy, Woolf encodes her developmental plot through characters who subtly reflect Clarissa's experience.[31] Perhaps most interesting of these is the infrequently noticed Rezia Warren Smith, wife of Clarissa's acknowledged double who has drawn critical attention away from the mirroring function of his wife. Rezia's life, like her name, is abbreviated in the novel, yet the course of her 'development' suggestively echoes that of the heroine. Like Clarissa, Rezia finds herself plucked by marriage from an Edenic female world with which she preserves no contact. Her memories highlight the exclusively female community of sisters collaboratively making hats in an Italian setting that is pastoral despite the surrounding urban context: 'For you should see the Milan gardens!' she later exclaims, when confronted with London's 'few ugly flowers stuck in pots!' (p. 34). The cultural shift from Italy to England, like the shift from Bourton to London, locates this idyllic female life in a distant, prelapsarian era – before the war, before industrialization, before marriage. Marriage and war explicitly coalesce for Rezia as agents of expulsion from this female paradise: Septimus comes to Milan as a British soldier and proposes to Rezia to alleviate his war-induced emotional anesthesia. Rezia's memories of Italy, a radiant temporal backdrop of her painful alienation in marriage and a foreign culture, provide a pointed parallel to Clarissa's memories of Bourton. And Rezia's final pastoral vision, inspired by the drug administered after Septimus's suicide, significantly begins with her sense of 'opening long windows, stepping out into some garden' (p. 227), thus echoing Clarissa's first recollection of Bourton, where she had 'burst open the French windows and plunged . . . into the open air' (p. 3). The death of her husband releases Rezia to return imaginatively to a past she implicitly shares with Clarissa: the female-centered world anterior to heterosexual bonds. After this moment of imaginative release and return Rezia disappears from the novel, having accomplished the function of delicately echoing the bifurcated structure of the heroine's development.

The relation of Clarissa and Rezia exists only for the reader; the two women know nothing of each other.[32] Woolf employs a different strategy for connecting Clarissa with Septimus, whose death severs the link between these female characters, releasing each to a new developmental stage, Rezia to return imaginatively to the past, Clarissa at last to transcend that past. Septimus's suicide enables Clarissa to resolve the developmental impasse that appears to be one cause of her weakened heart, her constricted vitality. Critics have amply explored Septimus's role as Clarissa's double. As important as this psychological doubling, however, is Woolf's revision of developmental plots, her decision to transfer to Septimus the death she originally imagined for Clarissa,[33] to sacrifice male to female development, to preserve her heroine from fictional tradition by substituting a hero for a heroine in the plot of violently thwarted development, a plot that has claimed such heroines as Catharine Linton, Maggie Tulliver, Emma Bovary, Anna Karenina, Tess Durbeyfield, Edna Pontellier, Lily Bart, and Antoinette Cosway Rochester. By making Septimus the hero of a sacrificial plot that enables the heroine's development, Woolf reverses narrative tradition.

It is a critical commonplace that Clarissa receives from Septimus a cathartic, vicarious experience of death that releases her to experience life's pleasures more deeply. Woolf's terms, however, are more precise. The passage describing Clarissa's reaction to Septimus's suicide suggests that he plays a specific role in Clarissa's emotional development. Woolf composes this passage as a subtle but extended parallel to Clarissa's earlier reminiscence of her love for Sally and Bourton.[34] The interplay between the language and structure of these two meditative interludes, the two major sites of the developmental plot, encodes Clarissa's exploration of a conflict more suppressed than resolved. By interpreting Septimus's suicide in her private language of passion and integrity. Clarissa uses the shock of death to probe her unresolved relation to her past. The suicide triggers Clarissa's recurrent preoccupation with this past, providing a perspective that enables her belatedly both to admit and to renounce its hold. On the day in June that encloses the action of *Mrs Dalloway*, Clarissa completes the developmental turn initiated thirty years before.

Woolf prepares the parallels between the two passages by inaugurating both with Clarissa's withdrawal from her customary social milieu. The emotions prompting Clarissa's first meditation on Sally and the past are initially triggered by her exclusion from Lady Bruton's lunch. Woolf then describes Clarissa's noontime retreat to her solitary attic room as a metaphorical departure from a party: 'She began to go slowly upstairs . . . as if she had left a party . . . had shut the door and gone out and stood alone, a single figure against the appalling night'; Clarissa is 'like a nun withdrawing' (p. 45). Later that night, when Clarissa hears the news of Septimus's suicide, she does leave her party and retreats to an empty little room where 'the party's splendor fell to the floor' (pp. 279–80). The first passage concludes with her preparations for the party, the second with her deliberate return to that party. Within these enclosed narrative and domestic spaces, Clarissa relives through memory the passionate scene with Sally on the terrace at Bourton. The second passage replays in its bifurcated structure the male intervention that curtails the original scene. In this final version of the female/male juxtaposition, however, the emotional valences are reversed.

Clarissa's meditation on Septimus's death modulates, through her association of passion with death, to a meditation on her relation to her past. Woolf orchestrates the verbal echoes of this passage to evoke with increasing clarity the scene with Sally Seton. Septimus's choice of a violent, early death elicits in Clarissa the notion of a central self preserved: 'A thing there was that mattered; a thing, wreathed about with chatter, defaced, obscured in her own life. . . . This he had preserved' (p. 280). The visual image of a vital, central 'thing' initiates the link with the earlier description of passion as 'something central which permeated' (p. 46). The echoes between these passages develop through their similar representations of passion's ebb: 'closeness drew apart; rapture faded, one was alone' (p. 281); 'But the close withdrew; the hard softened. It was over – the moment' (p. 47). As Clarissa implies that only death preserves the fading moment of passion, she prepares for her repetition of the *Othello* line that has signified her love for Sally Seton: 'If it were now to die, 'twere now to be most happy' (pp. 51, 281). The metaphor of treasure which precedes this explicit allusion to the scene with Sally further connects Clarissa's response to Septimus ('had he plunged holding his treasure?' she wonders) with her memory of Sally's kiss as 'a present . . . a diamond, something infinitely precious' (pp. 52–3). Septimus's death evokes in Clarissa the knowledge of what death saves and what she has lost; her grief is not for Septimus, but for herself. Woolf weaves the verbal web between the two passages to summon once again the crucial scene with Sally on the terrace at Bourton, to enable Clarissa to confront her loss. Clarissa's appreciation of this loss, at last fully present to her consciousness, crystallizes in the contrast that concludes this segment of the passage: 'She had schemed; she had pilfered. She was never wholly admirable. . . . And once she had walked on the terrace at Bourton' (p. 282).

With this naming of the original scene, Woolf abruptly terminates Clarissa's recollection, replaying with a brilliant stroke Peter Walsh's interruption, the sudden imposition of the granite wall. The masculine intervention this time, though, is enacted not by Peter but by Richard, and not as external imposition but as choice. Clarissa's unexpected thought of Richard abruptly and definitively terminates the memory of Sally, pivoting the scene from past to present, the mood from grief to joy: 'It was due to Richard; she had never been so happy' (p. 282). The dramatic and unexplained juxtaposition encapsulates the developmental plot and the dynamics of its central scenes. This final replay of the developmental turn, and final microcosm of Woolf's narrative method, however, represent the abrupt transition positively. The joy inspired by Clarissa's thought of Richard persists as she celebrates 'this having done with the triumphs of youth' (p. 282). Woolf does not fill in the gap splitting past from present, grief from joy. We can only speculate that Septimus's sacrificial gift includes a demonstration of Clarissa's alternatives: to preserve the intensity of passion through death, or to accept the changing offerings of life. By recalling to Clarissa the power of her past *and* the only method of eternalizing it, he enables her fully to acknowledge and renounce its hold, to embrace the imperfect pleasures of adulthood more completely. Through Septimus, Woolf recasts the developmental impasse in the general terms of progression or death. In the final act of the developmental plot, she qualifies her challenge to the notion of linear, forward growth.

Woolf signals the shift in Clarissa's orientation by concluding the interlude with Clarissa's reaction to the old lady across the way, an unnamed character who only functions in the novel as an object of Clarissa's awareness. The earlier meditative passage concludes with Clarissa's reflection in the looking glass; this one with an analogous reflection of a future identity. After Clarissa's thoughts shift from Sally and the past to Richard and the present, Woolf turns the angle of vision one notch further to open a perspective on the future. The old lady solemnly prepares for bed, but this intimation of a final repose, recalling Clarissa's earlier ruminations on her narrowing bed, carries no onus for the heroine, excited by the unexpected animation of the sky, the news of Septimus's suicide, the noise from the party in the adjacent room. Release, anticipation, pleasure in change, regardless of its consequences – these are Clarissa's dominant emotions. Her identification with Septimus and pleasure in his suicide indicate her own relief in turning from her past. The gulf between Clarissa and the unknown lady discloses the female intimacy forfeited to growth, yet Clarissa's willingness to contemplate an emblem of age instead of savoring a memory of youth suggests a positive commitment to development – not to any particular course, but to the process of change itself. The vision of the old lady simultaneously concludes the developmental plot and the depiction of Clarissa's consciousness; the rest of the narrative turns to Peter and Sally. The developmental theme resides in the interplay between two interludes in the sequence of the day.

Freud's comparison of the pre-Oedpial stage in women to the Minoan–Mycenean civilization behind that of classical Greece provides a metaphor for the course and textual status of Clarissa's development. It also suggests a broader historical analogue to female development, though not an analogue Freud himself pursues. Freud's psychoanalytic version of ontogeny recapitulating philogeny assumes a genderless (that is, implicitly masculine) norm: personal development repeats the historical progression from 'savage' to civilized races.[35] In *Mrs Dalloway* Woolf intimates more specifically that *female* development condenses one strand of human history, the progression from matriarchal to patriarchal culture implicit in Freud's archeological trope. Woolf's fascination during the years she was composing *Mrs Dalloway* with the works of Jane Harrison and the *Oresteia*, which traces precisely the evolution from Mycenean to Athenian culture, may have fostered this concern with the relation of gender to cultural evolution.[36] The developmental plot embedded in *Mrs Dalloway* traces the outline of a larger historical plot, detached in the novel from its chronological roots and endowed with an uncustomary moral charge.

Woolf assigns the action of *Mrs Dalloway* a precise date: 1923, shortly after the war that casts its shadow through the novel. Through the experience of Septimus Warren Smith and the descriptions of soldiers marching 'as if one will worked legs and arms uniformly, and life, with its varieties, its irreticences, had been laid under a pavement of monuments and wreaths and drugged into a stiff yet staring corpse by discipline' (pp. 76–7), she suggests that the military discipline intended both to manifest and cultivate manliness in fact instills rigor mortis in the living as well as the dead. For women, the masculine war is disruptive in a different way. Woolf's imagery and plot portray the world war as a vast historical counterpart to male intervention in female lives. In one pointed metaphor, the 'fingers' of the European war are so 'prying and insidious' that they smash a 'plaster cast of Ceres' (p. 129), goddess of fertility and mother love, reminder of the force and fragility of the primary female bond. Rezia's female world is shattered by the conjunction of marriage and war. The symbolic association of war with the developmental turn from feminine to masculine orientation will be more clearly marked in *To the Lighthouse*, bisected by the joint ravages of nature and war in the divisive central section. By conflating Mrs Ramsay's death with the violence of world war, Woolf splits the novel into disjunct portions presided over separately by the mother and the father.

In *Mrs Dalloway*, Woolf more subtly indicates the masculine tenor of postwar society. The youngest generation in this novel is almost exclusively, and boastfully, male: Sally Seton repeatedly declares her pride in her 'five great boys'; the Bradshaws have a son at Eton; 'Everyone in the room has six sons at Eton' (p. 289), Peter Walsh observes at Clarissa's party; Rezia Warren Smith mourns the loss of closeness with her sisters but craves a son who would resemble his father. Elizabeth Dalloway is the sole daughter, and she identifies more closely with her father than her mother (the plaster cast of Ceres has been shattered in the war). Male authority, partially incarnate in the relentless chiming of Big Ben, is more ominously embodied in the Doctors Holmes and Bradshaw, the modern officers of coercion. Septimus is the dramatic victim of this authority, but Lady Bradshaw's feminine concession is equally significant: 'Fifteen years ago she had gone under . . . there had been no scene, no snap; only the slow sinking, water-logged, of her will into his. Sweet was her smile, swift her submission' (p. 152). The loose connections Woolf suggests between World War I and a bolstered male authority lack all historical validity, but within the mythology created by the novel the war assumes a symbolic function dividing a pervasively masculine present from a mythically female past.

Critics frequently note the elegiac tone permeating *Mrs Dalloway*, a tone which allies the novel with the modernist preoccupation with the contrast between the present and the past.[37] Nostalgia in *Mrs Dalloway*, however, is for a specifically female presence and nurturance, drastically diminished in contemporary life. Woolf suggests this loss primarily in interludes that puncture the narrative, pointing to a loss inadequately recognized by the conventions of developmental tales. The most obvious of these interruptions, the solitary traveler's archetypal vision, loosely attached to Peter Walsh's dream, but transcending through its generic formulation the limits of private consciousness, is not, as Reuben Brower asserts, a 'beautiful passage . . . which could be detached with little loss', and which 'does not increase or enrich our knowledge of Peter or of anyone else in the book'.[38] Through its vivid representation of a transpersonal longing for a cosmic female/maternal/natural presence that might 'shower down from her magnificent hands compassion, comprehension, absolution' (p. 86), the dream/vision names the absence that haunts *Mrs Dalloway*. In the mundane present of the novel, the ancient image of the Goddess, source of life and death, dwindles to the elderly nurse sleeping next to Peter Walsh, as in another self-contained narrative interlude, the mythic figure of woman voicing nature's eternal, wordless rhythms contracts, in urban London, to a battered old beggar woman singing for coppers. The comprehensive, seductive, generative, female powers of the Goddess split, in the contemporary world, into the purely nurturant energy of Sally Seton and the social graces of the unmaternal Clarissa, clad as a hostess in a 'silver-green mermaid's dress' (p. 264). The loss of female integration and power, another echo of the smashed cast of Ceres, is finally suggested in the contrast between the sequence envisaged by the solitary traveler and the most intrusive narrative interlude, the lecture on Proportion and Conversion, where Woolf appears to denounce in her own voice the twin evils of contemporary civilization. Rather than a sign of artistic failure, this interruption calls attention to itself as a rhetorical as well as ideological antithesis to the solitary traveler's vision. Bradshaw's goddesses of Proportion and Conversion, who serve the ideals of imperialism and patriarchy, renouncing their status as creative female powers, are the contemporary counterpart to the ancient maternal deity, now accessible only in vision and dream. The historical vista intermittently inserted in *Mrs Dalloway* echoes the developmental progress of the heroine from a nurturing, pastoral, female world to an urban culture governed by men.

One last reverberation of the developmental plot takes as its subject female development in the altered contemporary world. Through the enigmatic figure of Elizabeth, Woolf examines the impact of the new historical context on the course of women's development. Almost the same age as her mother in the earliest recollected scenes at Bourton, Elizabeth has always lived in London; the country to her is an occasional treat she associates specifically with her father. Elizabeth feels a special closeness to her father, a noticeable alienation from her mother. The transition so implicitly traumatic for Clarissa has already been accomplished by her daughter. By structuring the adolescence of mother and daughter as inverse emotional configurations, Woolf reveals the shift that has occurred between these generations. As Clarissa vacillates between two men, while tacitly guarding her special bond with Sally, Elizabeth vacillates between two women, her mother and Miss Kilman, while preserving her special connection with her father. Elizabeth's presence at the final party manifests her independence from Miss Kilman; her impatience for the party to end reveals her differences from her mother. The last scene of the novel highlights Elizabeth's closeness with her father, whose sudden response to his daughter's beauty has drawn her instinctively to his side.

The opposing allegiances of daughter and mother reflect in part the kinds of female nurturance available to each. Elizabeth's relation with the grasping Miss Kilman is the modern counterpart to Clarissa's love for Sally Seton. Specific parallels mark the generational differences. Miss Kilman's possessive desire for Elizabeth parodies the lines that emblazon Clarissa's love for Sally: 'If it were now to die, 'twere now to be most happy' becomes, for Elizabeth's hungry tutor, 'if she could grasp her, if she could clasp her, if she could make her hers absolutely and forever and then die; that was all she wanted' (pp. 199–200). Sally walks with Clarissa on the terrace at Bourton; Miss Kilman takes Elizabeth to the Army and Navy Stores, a commercial setting that exemplifies the web of social and military ties. Miss Kilman, as her name implies, provides no asylum from this framework. Losing the female sanctuary, however, brings proportionate compensations: Elizabeth assumes she will have a profession, will play some active role in masculine society. Woolf does not evaluate this new developmental course, does not tally losses and gains. If she surrounds the past with an aureole, she points to the future in silence. She offers little access to Elizabeth's consciousness, insisting instead on her status as enigma – her Chinese eyes, 'blank, bright, with the staring incredible innocence of sculpture' (p. 206), her Oriental bearing, her 'inscrutable mystery' (p. 199). Undecipherable, Elizabeth is 'like a hyacinth, sheathed in glossy green, with buds just tinted, a hyacinth which has had no sun' (p. 186); her unfolding is unknown, unknowable. Through the figure of Elizabeth as unopened bud, Woolf encloses in her text the unwritten text of the next developmental narrative.

The silences that punctuate *Mrs Dalloway* reflect the interruptions and enigmas of female experience and ally the novel with a recent trend in feminist aesthetics. The paradoxical goal of representing women's absence from culture has fostered an emphasis on 'blank pages, gaps, borders, spaces and silence, holes in discourse' as the distinctive features of a self-consciously female writing.[39] Since narrative forms normally sanction the patterns of male experience, the woman novelist might signal her exclusion most succinctly by disrupting continuity, accentuating gaps between sequences. 'Can the female self be expressed through plot or must it be conceived in resistance to plot? Must it lodge "between the acts"?' asks Gillian Beer, the allusion to Woolf suggesting the persistence of this issue for a novelist concerned with the links of gender and genre.[40] In her next novel Woolf expands her discrete silence to a gaping hole at the center of her narrative, a hole that divides the action dramatically between two disjunct days. *To the Lighthouse* makes explicit many of the issues latent in *Mrs Dalloway*. The plot of female bonding, reshaped as the story of a woman's attempts to realize in art her love for an older woman, rises to the surface of the narrative; yet Lily's relationship with Mrs Ramsay is unrepresented in the emblem Lily fashions for the novel, the painting that manifests a daughter's love for her surrogate mother as a portrait of the mother with her son. Absence is pervasive in *To the Lighthouse*. The gaps in *Mrs Dalloway* are less conspicuous, yet they make vital and disturbing points about female experience and female plots. The fragmentary form of the developmental plot, where the patterns of experience and art intersect, conceals as insigificance a radical significance. The intervals between events, the stories untold can remain invisible in *Mrs Dalloway* – or they can emerge through a sudden shift of vision as the most absorbing features of Woolf's narrative.[41]

Notes

1. *A Writer's Diary* (London: The Hogarth Press, 1953), 28 November 1928, p. 139.

2. In *The Madwoman in the Attic: The Woman Writer and the Nineteenth-Century Literary Imagination* (New Haven: Yale University Press, 1979), SANDRA M. GILBERT and SUSAN GUBAR claim that 'women from Jane Austen and Mary Shelley to Emily Brontë and Emily Dickinson produced literary works that are in some sense palimpsestic, works whose surface designs conceal or obscure deeper, less accessible (and less socially acceptable) levels of meaning. Thus these authors managed the difficult task of achieving true female literary authority by simultaneously conforming to and subverting patriarchal literary standards' (p. 73).

3. 18 June 1923, entry in Woolf's holograph notebook dated variously from 9 November 1922, to 2 August 1923; cited by CHARLES G. HOFFMANN, 'From

short Story to Novel: The Manuscript Revisions of Virginia Woolf's *Mrs Dalloway*', *Modern Fiction Studies*, **14**, 2 (Summer 1968), 183.

4. 'Women Novelists', in *Women and Writing*, ed. Michèle Barrett (New York: Harcourt Brace Jovanovich, 1979), p. 71.

5. *A Room of One's Own* (New York: Harcourt Brace Jovanocich, 1957), pp. 76–7.

6. 'Literary Criticism', *Signs*, **1**, 2 (Winter 1975), 435.

7. On women novelists' dissatisfaction with the plot of romantic love, see NANCY K. MILLER, 'Emphasis Added: Plots and Plausibilities in Women's Fiction', *PMLA*, **96**, 1 (January 1981), 36–48, and MARIANNE HIRSCH, 'A Mother's Discourse: Incorporation and Repetition in *La Princesse de Clèves*', *Yale French Studies*, **62** (1981) 67–87. On the particular shift that took place in the early twentieth century, see ELLEN MOERS, *Literary Women* (Garden City: Doubleday, 1977), especially pp. 352–68; JANE LILIENFELD, 'Reentering Paradise: Cather, Colette, Woolf and Their Mothers', in *The Lost Tradition: Mothers and Daughters in Literature*, ed. Cathy N. Davidson and E. M. Broner (New York: Frederick Ungar, 1980), pp. 160–75; and LOUISE BERNIKOW, *Among Women* (New York: Crown Publishers, 1980), pp. 155–93. In '"Women Alone Stir My Imagination"; Lesbianism and the Cultural Tradition', *Signs*, **4**, 4 (Summer 1979), Blanche Wiesen Cook points out that 'were all things equal, 1928 might be remembered as a banner year for lesbian publishing' (p. 718).

8. In his essay 'Femininity', published the following year, Freud explicitly uses the metaphor of strata: 'A woman's identification with her mother allows us to distinguish two strata: the pre-Oedipus one which rests on her affectionate attachment to her mother and takes her as a model, and the later one from the Oedipus complex which seeks to get rid of her mother and take her place with her father.' The essay is reprinted to *Women and Analysis*, ed. Jean Strouse (New York: Grossman Publishers, 1974), p. 92.

9. Woolf's reaction to *Ulysses*, recorded in her journal entries in September and October 1922, suggests her interest in counteracting the perspective of this 'callow school boy, full of wits and powers, but so self-conscious and egotistical that he loses his head' (*The Diary of Virginia Woolf*, vol. 2: 1920–1924, ed. Anne Olivier Bell New York: Harcourt Brace Jovanovich, 1978), 6 September 1922, p. 199. For structural echoes of *Ulysses* in *Mrs Dalloway* see MARGARET CHURCH, 'Joycean Structure in *Jacob's Room* and *Mrs Dalloway*', *International Fiction Review*, **4**, 2 (July 1977), 101–9.

10. The essay on Jane Austen in *The Common Reader* (1925) incorporates a review Woolf wrote just after completing *Mrs Dalloway*. Woolf had also reviewed works by and about Austen in 1920 and 1922. In the *Common Reader* essay, Woolf tacitly assigns herself the role of Austen's heir by speculating that the novels Austen would have written in middle age would have manifested Woolf's aesthetic goals. The essay demonstrates a complex dialectic between Woolf's and Austen's concern with silence.

11. *The Diary of Virginia Woolf*, vol. 2: 1920–24, ed. Anne Olivier Bell (New York: Harcourt Brace Jovanovich, 1978), 30 August 1923, p. 263.

12. In '*Mrs Dalloway*: The Communion of Saints', *New Feminist Essays on Virginia Woolf*, ed. Jane Marcus (London: Macmillan; Lincoln: University of Nebraska Press, 1981), p. 126, SUZETTE A. HENKE points out that in the manuscript version of the novel Sally Seton clearly reciprocates Clarissa's love. Until

recently, Sally has been remarkably absent from critical commentary on *Mrs Dalloway*. Recent discussions include JUDITH McDANIEL, 'Lesbians and Literature', *Sinister Wisdom*, **1**, 2 (Fall 1976), 20–3; EMILY JENSEN, 'Clarissa Dalloway's Respectable Suicide', in *New Feminist Essays on Virginia Woolf*, ed. Jane Marcus. I am indebted to Tina Petrig, whose illuminating essay on female relationships in Woolf entitled ' – all sorts of flowers that had never been seen together before – ', first alerted me to the crucial role of Clarissa's relationship with Sally.

13. For the power and endurance of Woolf's relationships with women, see her letters, especially the letters to Violet Dickinson in volume I; JANE MARCUS, 'Thinking Back Through Our Mothers', and ELLEN HAWKES, 'Woolf's Magical Garden of Women', in *New Feminist Essays on Virginia Woolf*; Phyllis Rose, *Woman of Letters: A Life of Virginia Woolf* (New York: Oxford University Press, 1978), pp. 109–24; and JANE MARCUS, 'Virginia Woolf and Her Violin: Mothering, Madness and Music'. Although the relationship of Woolf's life to her fiction is much more pronounced in *To the Lighthouse*, there are quiet parallels between Woolf's biography and Clarissa's; the death of Clarissa's mother and sister cast Sally in the emotional role assumed by Vanessa Stephen, the primary nurturing figure throughout Woolf's life.

14. *Mrs Dalloway* (New York: Harcourt, Brace & World, 1927), pp. 52–3. This passage suggests an analogy between the wrapped-up present of Sally's love and the buried subplot of female bonds. All future references to *Mrs Dalloway* will be placed in parentheses in the text.

15. See for example, SIGMUND FREUD, 'Femininity', in *Women and Analysis*, ed. Jean Strouse, p. 98; HELENE DEUTSCH, 'Female Homosexuality', in *The Psycho-Analytic Reader: An Anthology of Essential Papers with Critical Introductions*, ed. Robert Fliess (New York: International Universities Press, 1948), pp. 208–30; ADRIENNE RICH, *Of Woman Born: Motherhood as Experience and Institution* (New York: W. W. Norton, 1976); CATHARINE STIMPSON, 'Zero Degree Deviancy', *Critical Inquiry*.

16. 'Some Psychical Consequences of the Anatomical Distinction Between the Sexes', in Strouse, *Women and Analysis*, p. 24; 'Female Sexuality', in Strouse, p. 42.

17. For an analysis of this scene as part of a pattern of interruption in *Mrs Dalloway*, see EMILY JENSEN, 'Clarissa Dalloway's Respectable Suicide', in *New Feminist Essays on Virginia Woolf*, ed. Jane Marcus. Jensen's point of view is similar to mine, though she does not adopt a psychoanalytic approach, and sees Clarissa's development in more purely negative terms than I.

18. 'Mrs Dalloway in Bond Street', in *Mrs Dalloway's Party*, ed. Stella McNichol (New York: Harcourt Brace Jovanovich, 1975), p. 27.

19. ELIZABETH JANEWAY suggests the resonance of this name in an essay entitled '*Who Is Sylvia?* On the Loss of Sexual Paradigms', *Signs*, **5**, 4 (Summer 1980), 573–89. She concludes the essay by asking, 'Who is Sylvia, whose name carries an edge of wilderness and a hint of unexplored memory? We do not know, but we will surely recognize her when she comes.'

20. The Hogarth Press began publication of Freud's *Collected Papers* in 1924; the first volume of the *Standard Edition*, translated by James Strachey and published in its entirety by the Hogarth Press, did not appear until 1948. Woolf's review entitled 'Freudian Fiction', in *Times Literary Supplement*, 25

March 1920, reveals that she was familiar with the essential of Freudian theory, though opposed to a simplistic application of the theory in fiction.

21. 'Female Sexuality', in Strouse, *Women and Analysis*, p. 42; 'Femininity', in Strouse, p. 78.

22. 'Femininity', p. 78.

23. For an analysis of this female sphere, see CARROLL SMITH-ROSENBERG, 'The Female World of Love and Ritual: Relations between Women in Nineteenth-Century America', *Signs*, **1**, 1 (Autumn 1975), 1–30. In *The Reproduction of Mothering: Psychoanalysis and the Sociology of Gender* (Berkeley: University of California Press, 1978), Nancy Chodorow argues that the pre-Oedipal orientation is not terminated by the Oedipus complex, but continues as a powerful influence throughout a woman's life, triggering repeated conflicts between allegiances to women and men.

24. 'Femininity', p. 77.

25. 'Femininity', p. 85; 'Female Sexuality', p. 43.

26. 'Female Sexuality', p. 43.

27. This significant fact is also noted by Elizabeth Janeway, 'On "Female Sexuality"', in Strouse, p. 60, and Sarah Kofman, 'The Narcissistic Woman: Freud and Girard', *Diacritics*, **10**, 3 (Fall 1980), 45.

28. 'Femininity', p. 87.

29. 'Femininity', p. 92.

30. PHYLLIS ROSE makes this point about Baron Marbot's *Memoirs* in *Woman of Letters: A Life of Virginia Woolf*, p. 144.

31. Catharine R. Stimpson implies a parallel between the coding of 'aberrant' sexuality in the works of Gertrude Stein and Woolf. See 'The Mind, the Body, and Gertrude Stein', *Critical Inquiry*, **3**, 3 (Spring 1977), 505.

32. Another parallel between these women is established through the Shakespearean allusions. The recurrent lines from *Cymbeline* associate Clarissa with Imogen; Rezia's name (Lucrezia) recalls Shakespeare's narrative poem, *The Rape of Lucrece*. The situations in these works are remarkably similar: in both men dispute one another's claims to possess the most chaste and beautiful of women, the dispute prompts one man to observe and/or test the virtue of the other's wife, and this encounter culminates in the real or pretended rape of the woman and eventually her actual or illusory death. The analogy between these Shakespearean heroines more closely allies Clarissa with Lucrezia in a realm external to but signaled by Woolf's text.

33. In her 'Introduction' to the Modern Library edition of *Mrs Dalloway* (1928), Woolf explains that 'in the first version Septimus, who later is intended to be her double, had no existence. . . . Mrs Dalloway was originally to kill herself or perhaps merely to die at the end of the party' (p. vi).

34. EMILY JENSEN also discusses the relationship between these passages in 'Clarissa Dalloway's Respectable Suicide', *New Feminist Essays*.

35. Freud claims, for example, that 'We can thus judge the so-called savage and semi-savage races; their psychic life assumes a peculiar interest for us, for we can recognize in their psychic life a well-preserved, early stage of our own

development' (*Totem and Taboo*, trans. A. A. Brill, New York: Random House, 1946, p. 3).

36. In *'The Years* as Greek Drama, Domestic Novel, and Gotterdämmerung', *Bulletin of the New Public Library*, **80**, 2 (Winter 1977), 276–301, JANE MARCUS discusses the influence of Jane Harrison's work on Woolf's fiction. She points out that Woolf's library contained a copy of Harrison's *Ancient Art and Ritual* (1918), inscribed to Woolf by the author on Christmas, 1923. In *'Mrs Dalloway*: The Communion of Saints', SUZETTE A. HENKE mentions that Woolf's notes for *Mrs Dalloway* are in a notebook that contains her earlier reflection on Aeschylus' *Choephoroi*. Woolf was reading Greek texts diligently in 1922 and 1923 in preparation for her essay 'On Not Knowing Greek'.

37. See for example, MARIA DI BATTISTA, *Virginia Woolf's Major Novels: The Fables of Anonymous* (New Haven: Yale University Press, 1980); PHYLLIS ROSE, *Woman of Letters*; J. HILLIS MILLER, 'Virginia Woolf's All Souls' Day: The Omniscient Narrator in *Mrs Dalloway'*, in *The Shaken Realist: Essays in Honour of Rederick J. Hoffman*, ed. Melvin J. Friedman and John Vickery (Baton Rouge: Louisiana State University Press, 1970), 100–127.

38. '"Something Central which Permeated". Virginia Woolf and Mrs Dalloway', in *The Fields of Light* (New York: Oxford University Press, 1951), p. 135. Brower also significantly omits Sally Seton from this summary of the novel's plot.

39. XAVIÈRE GAUTHIER, 'Is there Such a Thing as Women's Writing?' *New French Feminisms*, ed. Elaine Marks and Isabelle de Courtivron (Amherst: University of Massachusetts Press, 1980), p. 164. The whole project of *écriture féminine* stresses the importance of representing women's silence and absence.

40. 'Beyond Determination: George Eliot and Virginia Woolf', *Women Writing and Writing about Women*, ed. Mary Jacobus (London: Croom Helm, 1979), p. 80. Beer analyzes Woolf's resistance to plot in *The Waves*.

41. I would like to thank Marianne Hirsch, Elizabeth Langland, Diane Middlebrook, Marta Peixoto, Lisa Ruddick, Sanford Schwartz and Janet Silver for their helpful commentary on this essay.

6 'The Third Stroke': Reading Woolf with Freud*

MARY JACOBUS

Like the two previous essays, this one is psychoanalytic in its interests, but also literary in its psychoanalysis, as the piece crosses over between readings of Woolf and readings of Freud. Jacobus's attention may be directed to highly specific passages and coincidences of detail between the two authors, but she uses this precise attentiveness to make arguments about the relationship between myth and theory, and between fiction and history, which have broad applications to contemporary issues in feminist theory. Drawing at this point on the work of Julia Kristeva, as well as on *To the Lighthouse* and fantasmatic memories recorded in Freud and Woolf, Jacobus takes issue with those who have sought to claim for the original relations of mothers and daughters a simple purity only subsequently marred by the interventions of a culture seen as obtrusively masculine. Jacobus does not deny the power and the psychic necessity of such a myth, but insists that it should be seen as just that: a pre-historical, pre-mature story that can only acquire its significance from a point at which it must figure as what has been lost.

There is a joking saying that 'Love is homesickness.' Whenever a man dreams of a place or a country and says to himself, while he is still dreaming: 'this place is familiar to me, I've been here before', we may interpret the place as being his mother's genitals or her body.
(Sigmund Freud, 'The "Uncanny"', 1919)[1]

Characteristically, it is childhood memories that arouse powerful, retroactive feelings of loss. Yet childhood memories may also appear quite inconsequential. These are what Freud called 'screen memories' ('*Deckerinnerungen*') to indicate that they are never about what they seem. I want to use Freud's writing on screen memory as a way of reading Woolf's *To the Lighthouse* (which Woolf said would have 'father's

---

* Reprinted from Susan Sheridan (ed.), *Grafts: Feminist Cultural Criticism* (London: Verso, 1988), pp. 93–110.

character done complete in it; and mother's; and St Ives; and childhood').[2] Taking my cue from Toril Moi, who reminds us in her introduction to *Sexual/Textual Politics* (1985) that the Hogarth Press published the first English translations of Freud,[3] I'll try to suggest how reading Woolf with Freud might rescue her for the alternative feminist reading Moi proposes. Specifically, I want to reread Mrs Ramsay, whose maternal absence haunts the novel. Whether idealized as the androgynous artist or criticized as a meddling *Hausfrau*, she defines the artist-daughter, Lily Briscoe, in terms not of maternal plenitude but lack – 'Women can't write, women can't paint,' says Charles Tansley[4] – but artists can't mother, either. Lily's cry of longing ('"Mrs Ramsay! Mrs Ramsay!" she cried, feeling the old horror coming back – to want and want and not to have', p.229) expresses a horror on the far side of desire. Her compensatory vision at the end of the novel permits the completion of Woolf's design with a single line. One way to read this line would be as the 'third stroke', or the imaginary plenitude of the phallic mother. Another might be to see it as the line which at once fixed Lily and finished the book, placing both in a stabilizing, specular relation to artist and author, and thereby constituting them. Alternatively, one could see it as the line of minimal difference that makes possible the process which Kristeva calls abjection – the earliest emergence of the subject as distinct from the mother, and the entry of that not-yet-subject into signification.[5] I'll be exploring these readings in what follows, and in an attempt to sketch their significance for feminist psychoanalytic theory. In other words, as well as offering a partial rereading of *To the Lighthouse*, I want to suggest that there may be a specifically feminine dimension to the retrospective form of desire which Freud defines, in 'The Uncanny', as masculine homesickness for the material body.

The third stroke

Mrs Ramsay manifests herself in the novel as an absence. During her lifetime, in Part 1, she experiences herself as 'a wedge-shaped core of darkness, something invisible to others' (p. 72). After her death, which takes place in Part II, this 'core of darkness' or 'wedge of darkness' (p. 73) becomes the 'odd shaped triangular shadow' (p. 229) cast by an unseen object which brings Mrs Ramsay back to life and enables Lily Briscoe to complete her painting without the sitter. 'Mrs Ramsay – it was part of her perfect goodness to Lily – sat there quite simply, in the chair, flicked her needles to and fro, knitted her reddish-brown stockings, cast her shadow on the step. There she sat' (p. 230). Lily's final line is drawn in defiance of emptiness; in the gap between having and not having the mother. To this extent her design (like all signs in their denial of absence) is fetishistic. The completed picture stands as an imaginary, iconic finish to a novel whose governing tense is also the finished past perfect employed in the painful, parenthetical hiatus of part II ('Time Passes') which offhandedly narrates the death of Mrs Ramsay: '(Mr Ramsay stumbling along a passage stretched his arms out one dark morning, but, Mrs Ramsay having died rather suddenly the night before, he stretched his arms out. They remained empty)' (pp. 146–7). The syntax gives only to take away. Mr Ramsay's arms remain empty; a line stands for, or in place of, what is missing. Missing Mrs Ramsay is the theme of Lily's meditation as she works on her picture, years later.

Another name for this emptiness, this missing of the mother, might be nostalgia, or the longing to possess something one never had. Lily's cry of desolation ('to want and want and not to have', p. 229) recognizes the impossibility of assuaging desire. At the risk of ventriloquizing Woolf, I want to suggest that this horror of wanting, or privation – this lack symbolized by the absent mother – can be read as shadowing more than Woolf's novel, more than Lily Briscoe's status as woman artist. A related nostalgia shadows feminist criticism when it follows Woolf's injunction to think back through our mothers. Feminist revisions of psychoanalysis have tended to emphasize the project which Freud envisaged, archaeologically, as the excavation of the pre-Oedipal Minoan – Mycenaean civilization buried beneath the Hellenic Oedipal. In its desire to reconstitute femininity, not on the Freudian bedrock of penis envy, but on the imaginary plenitude symbolized by the maternal body, feminist theorizing about the mother risks creating its own myth – forgetting that the pre–Oedipal is the long shadow cast by the Oedipus myth, just as nostalgia is set in motion by imaginary deprivation. In his essay on screen memories, Freud refers to myths as a nation's 'childhood memories'. But perhaps childhood 'memories' are themselves really myths pressed into the service of theory (whether Freudian or feminist) – a speculation to which I will return.

In *To the Lighthouse*, Mrs Ramsay is at once the lost object that casts a shadow, and the source of light – both darkness and lighthouse. Mrs Ramsay identifies 'that stroke of the Lighthouse, the long steady stroke, the last of the three' as 'her stroke' ('this thing, the long steady stroke, was her stroke', p. 73). Is the long stroke an autoerotic caress, or a mark, like Lily Briscoe's line? Perhaps both; Gayatri Spivak, in an essay called 'Unmaking and Making in *To the Lighthouse*', suggests that Mrs Ramsay's knitting symbolizes both autoerotic activity and a form of feminine texuality.[6] Specularity and desire meet one another in the third stroke. 'She looked up over her knitting and met the third stroke and it seemed to her like her own eyes meeting her own eyes'; the hypnotic lighthouse began (so much her, yet so little her') seems to 'strok[e] with its silver fingers some sealed vessel in her brain whose bursting would flood her with delight' (p. 74). This imaginary maternal *jouissance* is set against the unravelling of identity, whether by time, death, or unconsciousness, in Part II. Formally, the novel pivots on itself towards just such an erotic self-completion, the 'wholeness not theirs in life' by which, Lily thinks, lovers create 'globed compacted things' (p. 218). The desire at work in the knitting of this text is that Lily's desire should be rounded on itself, like Mrs Ramsay's eyes meeting her own eyes – that lack should be completed (with the shadow of presence) or feminine deprivation veiled (with a web of words); or, at the end of the novel, the dissolution of language fixed by a specular image (a picture, a line).

The problem of disproportion which haunts Lily as she works on her picture might be re-expressed in Freudian terms as the problem of sexual difference. Faced with the absence of the mother's penis (so Freud's story goes), the little boy realizes, in retrospect, the reality of the threat of castration for himself, while the little girl at once perceives her deprivation.[7] One means of denying both sexual difference and castration anxiety is to imagine the mother as phallic or uncastrated, as possessing a third member. In this way, the boy protects himself from anxiety, while the girl is relieved of her sense of deprivation. Another solution – which may overlap with the first – is fetishism, whereby two contradictory ideas are simultaneously entertained. For the fetishist, the mother both is and is not castrated. The contradiction (again, in Freud's account) is often maintained by means of a screen, perhaps a garment or an object associated with the last moment before the mother's 'castrated' genitals are unveiled.[8] In the opening scene of *To the Lighthouse*, Mrs Ramsay makes 'some little twist of the reddish-brown stocking she was knitting' (p. 7), defending her son James against the castrating presence of his father ('lean as a knife, narrow as the blade of one', p. 6) when he declares, with remorseless rationality, 'It won't be fine', thereby delaying the trip to the lighthouse for a decade. To Lily, at work on her picture, relations between the sexes represent the threat of castration as well as complementarity ('that razor edge of balance between two opposite forces', p. 219). I want to begin to unravel both Mrs Ramsay's stocking and Woolf's web of words by exploring the psychoanalytic representation of sexual difference in Freud's writing on screen memory – reading Woolf 'with' Freud, but also 'beyond' him, to an attempted reformulation of the mother's role in representation that is based on Kristeva's revisionary reading of the pre-Oedipal.

Virginia Woolf

'The story of M'

In *The Psychopathology of Everyday Life* (1901), Freud devotes a chapter to
the topic of 'Childhood Memories and Screen Memories'. By 'screen
memories', he means memories which preserve something that is
'screened off' or unavailable to consciousness. This associative link
always involves a chronological distortion or displacement; childhood
memories are either retroactively recovered or retrospectively
constructed, or else a screen memory may hide something contiguous to
it in time. Either way, the status of memory is put in question. Instead of
being a recovery of the past in the present, it always involves a revision,
reinscription, or re-presentation of the past, which ceases to be its
referent. Memory becomes a mode of 'reproduction' (Freud's term)
which resembles dreaming not only in its distortions and displacements,
but in its paradoxical relation to an unconscious forgetting that is always
purposive. Just as dreaming represents the fulfilment of a repressed
wish, memory represents a paradoxical desire – the wish to forget.

Freud tells us that he was first alerted to the 'tendentiousness' (his
phrase) of memory by the striking fact that memories preserved from
earliest childhood appear to be, as he put it, 'indifferent' (in German,
both *indifferenten and gleichgultigen*) – a word which recurs insistently in
his writing about screen memory. What is an 'indifferent' memory?
Could it be one in which the inscription of sexual difference has been
elided or repressed? As if to prove the point, Freud instances a single,
telling example, an anecdote which turns on the displacement of sexual
difference on to the ostensibly indifferent scene of writing. Here is the
story:

> A man of twenty-four has preserved the following picture from his
> fifth year. He is sitting in the garden of a summer villa, on a small chair
> beside his aunt, who is trying to teach him the letters of the alphabet.
> He is in difficulties over the difference between *m* and *n* and he asks
> his aunt to tell him how to know one from the other. His aunt points
> out to him that the *m* has a whole piece more than the *n* – the third
> stroke. There appeared to be no reason for challenging the
> trustworthiness of this childhood memory: it had, however, only
> acquired its meaning at a later date, when it showed itself suited to
> represent symbolically another of the boy's curiosities. For just as at
> that time he wanted to know the difference between *m* and *n*, so later
> he was anxious to find out the difference between boys and girls, and
> would have been very willing for this particular aunt to be the one to
> teach him. He also discovered then that the difference was a similar
> one – that a boy, too, has a whole piece more than a girl; and at the

time when he acquired this piece of knowledge he called up the recollection of the parellel curiosity of his childhood.[9]

In this classic representation of sexual difference ('The m has a whole piece more than the n'), binary opposition leads to asymmetry. M has something that n lacks, 'the third stroke – 'a whole piece more'; though to be without it is to be less than whole. That this asymmetrical structure – the minimal difference between two adjacent characters which translates into feminine lack – determines the entire phallocentricity of the Freudian inscription of sexual difference hardly needs underlining. Figuring sexual difference as a hierarchy of plus or minus reveals the castrating cost for femininity of 'the third stroke'. This is the stroke that Woolf gives back to Mrs Ramsay in an act of reparation complicit with the mother's own wishful attempt to screen her son from the castrating Oedipal father in the novel's opening scene. With Freud's scenario in mind, we can see that Lily's solution to the puzzle of the sexes (her conversion of disproportion into complementarity) is continuous with the fetishistic solution. Both involve a denial of sexual difference, or a veiling of castration.

But why the imbrication of sexual difference in the scene of writing/ knitting? *La plume de ma tante*, in Lacanian terms at any rate, would have it that all subjects are 'castrated', whether *avant* or *à la lettre*, irrespective of anatomical gender. When it comes to the phallic signifier, lack is the only lesson taught. The subtext of the aunt's lesson is as inexorable as Mr Ramsay's castrating prediction about the weather. Freud's story implies that the difference between m and n only becomes significant (i.e. ceases to be 'indifferent') for the boy retrospectively, when his sexual curiosity and desires have become aroused and focused on his aunt (an obvious mother substitute). But to write is already to have acceded to the inscription of sexual difference – to the Oedipus complex, and to the anxieties about loss of wholeness which Freud calls castration anxiety. Lacan has it, famously, that 'the unconscious is structured like a language', always unravelling essence and identity by its differential, metonymic movement, which is the movement of desire-in-language. Writing as such can never be regarded as 'indifferent', any more than screen memories. Rather, inscription screens not only the differentiating ban on desire for the mother (the Oedipal moment elaborated by Freud) but the traumatic discovery of the mother's 'castration', which for Lacan, in his essay on 'The Meaning of the Phallus', provides the founding moment of sexual difference.[10]

Freud's overdetermined example turns out to be an exemplary 'mnemonic' for the inscription of sexual difference. The swift reconsolidation of masculine identity occurs by reference to a woman (an aunt) who stands in for the always absent mother. The aunt lacks what the boy wants; teaching him his own desire, she becomes the phallus for him. If we follow Freud's trajectory in *The Psychopathology of Everyday Life*, we find that his later revisions to this chapter initiate a series of autobiographical memories which uncover his own earliest memories of his mother. These concern nothing less than childhood sexuality, which for Freud is always Oedipal rather than pre-Oedipal; hence, we might say that in his case, too, childhood memories screen (and so preserve) the repressive inscription of sexual difference. Preceding the 'indifferent' scene of writing by many years, these autobiographical recollections are of particular significance for Freud himself, since they lead (in the correspondence of 1897 with Fliess, on which they are based) to the momentous 'discovery' of the Oedipus complex and to his entire theory of sexuality. Screen memories involving the mother lead Freud to the founding myth of psychoanalysis.

The dandelion phantasy

At this point, however, my concern is not so much with the Oedipal as with the pre-Oedipal, and with the daughter whose inscription under the letter '*n*' renders her history one of hypothetical deprivation. Since considerations of femininity and the pre-Oedipal have become almost inseparable in current feminist psychoanalytic theory, I want now to explore this mythic prehistory – with its feminine nostalgia for maternal origins and its retroactive excavation of the buried civilization beneath the Hellenic myth – by way of Freud's other extended discussion of screen memory, in an essay of 1899. Under the cloak of an assumed persona, he constructs a first-person narrative that is generally assumed to be autobiographical. His narrative focuses on a remembered scene, or *mise en scène*, which 'appears to [him] fairly indifferent' yet which has become fixed in his memory. Once more he is concerned to show that 'precisely what is important is repressed and what is indifferent retained' (p. 306).

Here is the screen memory which Freud refers to as 'the dandelion phantasy':

> I see a rectangular, rather steeply sloping piece of meadow-land, green and thickly grown; in the green there are a great number of yellow flowers – evidently common dandelions. At the top end of the meadow there is a cottage and in front of the cottage door two women are standing chatting busily, a peasant-woman with a handkerchief on her head and a children's nurse. Three children are playing in the grass. One of them is myself (between the age of two and three); the two others are my boy cousin, who is a year older than me, and his

sister, who is almost exactly the same age as I am. We are picking the yellow flowers and each of us is holding a bunch of flowers we have already picked. The little girl has the best bunch; and, as though by mutual agreement, we – the two boys – fall on her and snatch away her flowers. She runs up the meadow in tears and as a consolation the peasant-woman gives her a big piece of black bread. Hardly have we seen this than we throw the flowers away, hurry to the cottage, and ask to be given some bread too. And we are in fact given some; the peasant-woman cuts the loaf with a long knife. In my memory the bread tastes quite delicious – and at that point the scene breaks off.[11]

At the outset, Freud confesses himself baffled. Is the emphasis on the little boys' 'disagreeable behaviour to the little girl'? Is it on the yellow colour of the dandelions or the delicious taste of the bread? Colour and taste seem exaggerated, almost hallucinatory. Two episodes, it turns out have been grafted together. The 'memory' first emerged at the age of seventeen, when Freud returned to the country where he had spent his early childhood and experienced his 'first calf-love'[12] for a fifteen-year-old girl who was wearing a yellow dress when they met. He is indifferent to her now, as he is to dandelions. But was the colour that of dandelions? No; another flower – found in the Alps, but darker yellow – 'would exactly agree in colour with the dress of the girl I was so fond of'. The alpine reference dates the emergence of the 'screen memory' as belonging to a still later period, placing on it the time-stamp of his first acquaintance with the Alps where, as a young medical student, he again meets his two boy and girl cousins. This time Freud didn't fall in love; instead the fantasy is projected onto his father and uncle, who, Freud believes, had hopes that he and his cousin might marry and settle down in practical life. Not until later does Freud reflect that the father had meant well in this plan for his son. The 'memory', then, was actually created at the time of Freud's struggles to establish himself as 'a newly-fledged man of science':[13] indeed, it resonates with his current professional anxieties at the time of writing.

Though the scene itself may be genuine, it is charged with impressions and thoughts from a later date. Its hallucinatory vividness comes from this freight of unconscious desire. But Freud's analysis does not stop here. The yellow flowers remain to be explained as 'a representation of love'. 'Taking the flowers away from a girl means to deflower her.'[14] This (says Freud) is a scene of rape – a bold unconscious fantasy beneath the bashfulness of the seventeen-year-old and the rebellious indifference of the student. As Freud puts it, 'the coarsely sensual element in the phantasy' finds its way 'allusively and under a flowery disguise into a childhood scene'.[15] Representation – 'screen memory' – allows the repressed thought to become conscious, albeit in a censored form. One

final question remains to be answered. Is the whole 'dandelion phantasy'
only a retrospective construction? No, Freud allows as how it's genuine –
'there is a memory trace the content of which offers the phantasy a point
of contact'.[16] A prior inscription has retroactively acquired the
superscription which accounts for its vividness. 'In your case' (Freud tells
himself) 'the childhood scene seems only to have had some of its lines
engraved more deeply'.[17] As proof of the genuineness of the scene, he
adduces its unexplained residue, the two little boys, the peasant-woman,
and the nurse.

But there is another residue of 'the dandelion phantasy' – its
unexplored and (by Freud at least) unacknowledged pathos. Freud's
recollections concern himself as son. By the time of his self-analysis, he
was also a father. The (rejected) counter-movement of his own
psychoanalytic thinking at this period involved the allegations of his
women patients that they had been seduced by the fathers whom he
subsequently came to view as the object of their unconscious desires.
Giving up the 'seduction theory' as an explanation for hysteria made the
crucial break between the so-called 'science' of psychoanalysis and the
empirical or positivistic model traditionally associated with science. For
disillusioned Freudians like Jeffrey Masson, the editor of the recently
published Fliess correspondence,[18] this break constitutes an act of bad
faith on the part of a Freud anxious to distance himself from the
disturbing actualities of medical malpractice (Emma Eckstein's botched
nose job) and the historical reality of nineteenth-century child abuse. But
this is to lose sight of the liberating effects of psychoanalytic theory –
effects that are especially important, though also troubling, for
psychoanalytic feminism. The rejection of unmediated, one-to-one causal
connections between the social and the psychic (between the
contingencies of personal, family, and national history on one hand, and
the unconscious on the other) is a central, radical aspect of Freudian
theory. Its advantage for feminism lies in refusing socially determined, as
much as biologically determined, definitions of the gendered subject.
Instead of reinstituting the literal (like Masson), I'd like to stay with the
'myths' or prior inscriptions of psychoanalysis. The traces that underlie
memory, after all, carry their own freight of sexual desire as well as
ideology. It is not so much a matter of psychoanalysis ignoring history,
as a matter of acknowledging that Freud's choice of the Oedipus story as
his founding myth is itself historically determined.

'Proserpine gathering flowers'

Freudian psychoanalysis is often said to have a 'blind spot' in relation to femininity. Freud's appropriation of Greek myth has a specifically Oedipal blindness. Nineteenth-century Hellenism – the cultural determinant of Freud's psychoanalytic myth-making – reveals the hidden sexual biases of cultural history. In a letter to Fliess of 31 May 1897 (the same period as his 'Oedipal' dreams about his mother), Freud records a dream about having 'overaffectionate feelings for Mathilde', his daughter, 'only she was called Hella; and afterward I again saw 'Hella' before me, printed in heavy type'.[19] Hella was the name of Freud's American niece (and, incidentally, of a Germanic mother goddess). But his daughter Mathilde, Freud tells us, might also be called 'Hella' because 'she recently shed bitter tears at the defeats of the Greeks. She is enthralled by the mythology of ancient Hellas.' A young Greek girl, she is in love with the past of Greek myth. For Freud, the dream 'shows the fulfilment of my wish to catch a *Pater* as the originator of neurosis'.[20] One could gloss this as saying that Freud's dream fulfils his wish to find the seduction theory true (even at the cost of accusing himself), despite the fact that at this point he was on the verge of letting it go in order to found psychoanalysis on myth rather than history.

But what about his allusion to 'the mythology of ancient Hellas'? – the mythology which came so readily to hand soon afterwards when Freud moved via his memories of his mother to the conviction of the universality of the Oedipus complex and his abandonment of the seduction theory? Freud's denial of the role of the parent as seducer and his corresponding emphasis on the unconscious desires of the child privilege one myth (the Oedipal) over another (let's call it the pre-Oedipal). The residual pathos of the dandelion phantasy suggests that this memory screens a forgotten myth. What but the rape of Proserpine by Pluto, so calculated to make those yellow flowers bloom with hallucinatory brightness and to account for the archetypal violence and sacrificial pathos of the scene? 'As though by mutual agreement, we – the two boys – fall on her and snatch away her flowers. She runs up the meadow in tears.' Ovid's *Metamorphoses*, Book V, is brief enough to quote as an illustration of the powerful literary motif which gives the scene its resonance and provides its more-than-contingent residue – its glimpse of the loss of pre-Oedipal innocence, not as it affects sons, but, specifically, as it affects daughters:

There spring is everlasting. Within this grove Proserpine was playing, and gathering violets or white lilies. And while with girlish eagerness she was filling her basket and her bosom, and striving to surpass her mates in gathering, almost in one act did Pluto see and love and carry

111

her away: so precipitate was his love. The terrified girl called
plaintively on her mother and her companions, but more often upon
her mother. And since she had torn her garment at its upper edge, the
flowers which she had gathered fell out of her loosened tunic; and
such was the innocence of her girlish years the loss of her flowers even
at such a time aroused new grief.[21]

And who sees her gathering flowers? 'Her father's brother', Pluto. For
'father's brother' (uncle), we have to read 'father' in at least one of
Freud's *Studies in Hysteria* from the 1890s.[22] This is not just to say that
Freud – whether consciously or unconsciously – 'screens' the father's
part in the dandelion phantasy; but rather that all appropriations of myth
in a theoretical context are bound to be, in Freud's own terms,
'tendentious' or expropriative.

Selecting the Oedipus myth as his vantage-point for undoing the
untheorized empiricism of psychiatric medicine, Freud simultaneously
writes into his theory the sexual politics which consigns the pre-Oedipal
to forgetfulness. His inability to read the dandelion phantasy as a screen
for paternal desire is symptomatic of the masculinist blindness which has
rendered (and still renders) Freudian psychoanalysis suspect to many
feminists. A sociological reading of Freud would want to see patriarchal
rapine as the hidden but accurate model of relations between the sexes,
with incest as the unacknowledged, repressed 'norm' of the patriarchal
family. But my intention is not to rewrite psychoanalysis as an account of
the oppressive functioning of gender relations in patriarchal culture
(though it isn't, either, to privilege the myths of Hellenic culture as
culturally and historically universal). Rather, what intrigues me is the
way in which literature, whether forgotten or remembered, provides the
pre-text for psychoanalytic theory, both Freudian or feminist. In the
humanist high culture of nineteenth-century Germany, for instance,
Sophoclean tragedy ranks above Ovidean narrative. In this implicit
literary hierarchy, men are aligned with Greek tragedy (a 'strong' form
that normalizes violent contingency as moral justice) while women are
aligned with Roman pathos (a 'weak' form which makes passive
suffering aesthetically acceptable). My question then becomes: has
feminist criticism been coloured by this pathos in adopting the rape of
Proserpine as its pervasive matrilinear fiction?

The story that Freud 'forgets' – the myth which the Oedipus myth
screens – is the version of the pre-Oedipal which informs at least one
influential brand of feminist theory. This is the archetypal story of lost
mother-daughter relations, and the compulsory institution of patriarchal
heterosexuality which attends it – the rape of the girl child (Proserpine/
Korê) from her mother (Ceres/Demeter) by the patriarchal father (Dis/
Pluto). The rape of Proserpine could be read as the dark side of the

Oedipus complex; on this level, it is the subtext of Freud's own story about the girl's disappointed rejection of her mother for failing to equip her with a penis, and the forcible transfer of love from her first maternal object to a heterosexual love-object by way of incestuous desire for the father. in *Of Woman Born: Motherhood as Experience and Institution*, Adrienne Rich calls this rupture of mother-daughter relations in the interests of patriarchal heterosexuality 'the great unwritten story' the knowledge, before sisterhood, of 'mother-and-daughterhood'.[23] Rich's history of bodily expropriation is underwritten by the myth of the mother's mourning for her lost daughter and the daughter's mourning for the loss of a nurturing mother. Though there may be strategic effectiveness in Rich's use of the myth (and, before her, Phyllis Chesler's in *Women and Madness*),[24] I want to question not only its nostalgia, but its implicit psychic utopianism. Even as they indict the institutional effects of patriarchal oppression on women's minds and bodies, Rich and Chesler imply that relations to the body could be unalienated and psyches undivided. For both, bodily and psychic – hence, political – power would come by way of the fantasy which Freud calls 'the phallic mother', the woman who possesses the phallus (the story of '*m*'). Paradoxically, the mother-centred feminist narrative developed as an alternative to the oedipal narrative of psychoanalysis risks reinscribing a fiction which defends against castration anxiety at the price of denying sexual difference. If the mother has it, then there is only masculinity after all.

Expropriation and nostalgia

Is there no way for feminists to answer Woolf's call to 'think back through our mothers' without mimicking the fetishist's refusal of sexual difference – no way to read the myth of Proserpine, as it were, 'beyond Oedipus'? In an essay called 'Beyond Oedipus: The Specimen Story of Psychoanalysis', Shoshana Felman reads the Sophoclean story not as 'a myth of the possession of a story, but the myth . . . of the dispossession by the story';[25] psychoanalysis depends on a generative, fictive moment, when myth breaks with its origins to become theory (Felman calls this moment the 'expropriation' of myth). The myth of Proserpine and Ceres (Korê and Demeter) – itself a story of rape or expropriation – could be read in the same way as a myth of alienation between the myth and itself, or the breach between the theory and the fiction on which it is founded. This is the internal difference which Mrs Ramsay experiences in relation to the Lighthouse ('so much her, yet so little her'). But I don't want to lose sight of the symptomatic nostalgia of the myth in its feminist appropriation ('expropriation'). Nostalgia, one could say, screens the fiction of the pre-Oedipal. To put it another way, there never was a prior time for the subject, except as the Oedipal defines it retroactively. The mother is always already structured as division by the Oedipal; there is no violent separation because separation is inscribed from the start. Undermining the fiction of the pre-Oedipal, this reading points to something that may be symptomatic not about the myth, but about its nostalgic colouring.

In her recent book, *Reading Lacan* (1985), Jane Gallop explores the significance of the word 'nostalgia' in Lacan's classic essay, 'The Meaning of the Phallus'. Lacan makes the castration complex for both boys and girls turn on the perception of the mother's castration (rather than in response to paternal prohibitions on incest); Lacan's is thus both a more even-handed account than Freud's and one that is more easily expropriated for feminism. Only in the light of this perceived lack does castration take on its retrospective meaning for boys and penis-envy for girls. The distinction usually made is between threat on one hand, and deprivation on the other. What is unexpected in Lacan's account is the substitution of the term 'nostalgia' for the deprivation associated with the operation of the castration complex on women. As Gallop puts it, paraphrasing Lacan, 'Man's desire will henceforth be linked by law to a menace; but woman's desire will legally cohabit with *nostalgia*.'[26] Gallop reads Lacan's essay autobiographically, in the light of Lacan's own nostalgia for the earlier moment in psychoanalytic theory which is the starting point for his discussion of sexual difference – the 'great debate' of 1928–32 over feminine sexuality. In this doubling of autobiographical and theoretical concerns, she finds something that takes Lacan not so much 'beyond Oedipus' (as in Felman's reading of Freud) as beyond the theoretical stalemate over penis-envy to a redefinition of feminine desire.

For Gallop, nostalgia is a form of *'Nachträglichkeit'* – a regret for a lost past that occurs as a result of a present view of that past moment'. The nostalgia of penis-envy is similarly retrospective; it 'does not simply accompany the moment of castration, but rather is a retroactive effect . . . It is not that the girl experiences loss but rather that, looking back . . . she feels regret.'[27] Remembering (what has never been lost) is constitutive of nostalgia, just as one might say that the Oedipal is constitutive of the pre-Oedipal. Freudian theory says that the perception of loss on the part of the girl is perceived loss of what the boy (still) has, the phallus; feminist theory (in turn and in reaction) has redefined this retrospective perception of loss as mourning for the lost mother. Gallop, however, points the way to a different formulation. In French, she reminds us, *'Nostalgie'* is defined as (1) haunting regret for one's native land, or homesickness; and (2) melancholy regret, or unsatisfied desire. Freud claims that, psychoanalytically speaking, homesickness is longing to return to the lost home (womb) of the mother (sense 1). But unless she is imagined as the phallic mother, the mother is always lost, the subject forever abroad. As Gallop puts it, 'the discovery that the mother does not have the phallus means that . . . the subject is hence in a foreign land, alienated,' For Lacan, desire itself is the offshoot of a need which 'finds itself alienated'; by definition, desire is unsatisfied (sense 2).[28] The alienated need articulated as nostalgia is desire for the mother as 'grounding for the subject' – desire for undifferentiated wholeness. But there is no subjectivity without division. Nostalgia, then, might be called the feminine articulation of what it means to become a gendered subject – a crucial admission that castration may bear differently on women while yet structuring all subjects as necessarily constituted in, and by, lack and division.

Wrapping things up

In *Of Woman Born*, Rich calls *To the Lighthouse* 'the most complex and passionate vision of mother–daughter schism in modern literature . . . one of the very few literary documents in which a woman has portrayed her mother as a central figure'.[29] Rich turns to Woolf's novel immediately after her autobiographical account of her own ambivalent relation to her gifted but family-centred mother; she cites Lily Briscoe, her head on Mrs Ramsay's lap, as desiring (in Woolf's words) 'not knowledge but unity . . . not inscription on tablets . . . but intimacy itself'.[30] In Rich's account of the novel, Mrs Ramsay's final unavailability to Lily is offset by Lily's attainment of independence in her art – her acceptance that only in aesthetic inscription (in her case, painting) lies the possibility for self-differentiation. (It may be worth noting in this context that Rich's own mother was called Helen – a name as coloured by masculine desire as the maternal landscape of Freud's 'Uncanny'; and for Lily also, Mr Ramsay has too much possessed his wife.) Calling 'the loss of the daughter to the mother, the mother to the daughter . . . the essential female tragedy',[31] Rich introduces the Eleusinian mysteries as a compensatory fantasy of the mother's miraculous recovery in the mythic prehistory of women. Division engenders desire, desire engenders a retroactive fiction of unmediated mother–daughter relations whose sign is nostalgia.

At this point I want to wrap things up by imbricating fetishism, nostalgia, and the self-differentiation which makes signification possible in Kristeva's account of the mother in *Powers of Horror* (1980) and her later essay, 'L'abjet d'amour' from *Histoires d'amour* (1983).[32] In *To the Lighthouse*, Mrs Ramsay goes upstairs to find her youngest children, Cam and James, wide awake in their beds, quarrelling over 'that horrid skull again . . . Cam couldn't go to sleep with it in the room, and James screamed if she touched it' (p. 131). Here the skull of a wild boar ('"only an old pig . . . a nice black pig like the pigs at the farm"', p. 132) – is central to a scene of sexual differentiation. This is the sight of something (something naked as death or castration – a threatening absence) to which boy and girl react differently in Freud's account. To Cam, the girl, the tusked skull is terrifying, so Mrs Ramsay wraps it in her shawl, 'and wound it round the skull, round and round and round', weaving a rhythmic web of words which lull her to sleep ('it was like a bird's nest; it was like a beautiful mountain', and so on). For James the skull must be left intact – 'see, she said, the boar's head was still there; they had not touched it . . . it was there quite unhurt' (pp. 132–3). Mrs Ramsay veils the phallic sight in semiotic babble (so to speak), while reassuring James that it's still there, underneath – a fetishistic solution to the horror of wanting.

Gayatri Spivak sees this double gesture as a Derridean moment of undecidability.[33] But a memory thrown up by Freud's sequence of dreams about his mother (the dreams on which he draws in both his letters to Fliess and in 'Screen Memory') suggests a reading that turns the undecidability of fetishism towards what Kristeva calls 'abjection'. One of Freud's dream-memories concerns his elderly nurse – not only, he claims, his earliest seducer (his 'prime originator'), but the woman who first made him sexually ashamed: 'she was my teacher in sexual matters and complained because I was clumsy and unable to do anything', he told Fliess.[34] His dream goes like this: 'At the same time I saw the skull of a small animal and in the dream I thought "pig!"' Freud goes on to associate the dream with Fliess's wish that he might find, as Goethe once did, a skull on the Lido to enlighten him. 'But I did not find it. So [I was] a "little blockhead" [literally 'a little sheep's head'].'[35] (The allusion here is to Goethe's story of finding 'the split skull of a sheep . . . which gave [him] the idea for the so-called "vertebral" theory of the skull'; the skull crops up elsewhere in a dream which Freud analysed, like this one, in terms of his professional anxieties).[36]

In his self-interpretation, Freud traces back his desire to distinguish himself as a therapist not just to childhood clumsiness, but to a moment of denigration ('Pig!' or 'little blockhead') that involves a mother substitute – one who (we know) happened to be quite literally 'abjected', or sacked and imprisoned during Freud's infancy for petty theft, and from little Sigmund himself. He goes on to confess to identification with his thieving nanny. But what if we were to see the abjectly denigrated 'Pig!' as initiating the movement which leads not just to the old nanny's symbolic 'abjection' as a classic betrayer of infant and family trust, but to the child's abjection of the mother in the interests of self-differentiation? Then we would have something very like the paradigm by which, for Kristeva, the earliest separation from the maternal body takes place. Horror ('that horrid skull') or disgust (inability to manage one's sexual or natural functions – 'Pig!') leads to the demarcation of boundaries, that first mapping of the body by the mother's care through which the not-yet-subject enters into signification. If the skull is a death's head, then what has become visible is the corporality of a distinct body, that which threatens identity and undoes the border between the child's body and its mother's. Like death infecting life, the skull disturbs identity and order. It also stands in the place of the mother's loss, which is her lack or deficiency in one important respect. 'All abjection', writes Kristeva, 'is in fact recognition of the *want* on which any being, meaning or desire is founded.'[37] The 'horror of wanting' experienced by Lily before the empty steps repeats that inaugural loss, the want on which subjectivity itself is founded in the first instance.

In Kristevan terms, the hallucinatory moment which restores Mrs Ramsay to Lily ('there she sat') combines fetishism and abjection. 'For the absent object, there is a sign. For the desire of that want, there is a visual hallucination.'[38] The corollary of visual hallucination is voyeurism – that phobic looking in quest of the horrifying absence which is the origin of (matro)phobia. 'Fifty pairs of eyes were not enough to get round that one woman with', thinks Lily (p. 224). Signs themselves are a kind of fetish; as Kristeva puts it, 'The writer is a phobic who succeeds in metaphorizing in order to keep from being frightened to death'.[39] Recalling the incantation with which Mrs Ramsay veils the skull, we might say that Woolf herself, in Part II of *To the Lighthouse* ('Time Passes') metaphorizes loss of identity, or mortality, as the passage of time – the gradual decay of a house – in order to keep from being frightened to death; signs stand in for the horror of absence when, in Kristeva's own evocative phrase, 'death brushes [her] by'. In this central section of Woolf's novel, there is no subject of consciousness; the impersonal 'Nothing' (p. 143) becomes the non-subject or not-yet-subject, the 'abject'. 'Analysis', writes Kristeva, 'give[s] back a memory, hence a language, to the unnameable and nameable states of fear.'[40] Lily wonders in the face of the empty steps: 'how could one express in words these emotions of the body? express that emptiness there?' (p. 202). Kristeva would name that emptiness 'fear' – 'The *void* upon which rests the play with the signifier' and 'the arbitrariness of that play', she writes, 'are the truest equivalents of fear'.[41] In *To the Lighthouse*, language becomes what Kristeva calls 'our ultimate and inseparable fetish' (Kristeva, p. 37), and signs become the object of phobic desire, in a despairing attempt to fill the void figured by the loss of the maternal. And yet it is only through identifying with that void that Lily Briscoe finally comes into being as an artist.

During the traumatic central interlude of *To the Lighthouse*, not only does identity unravel, but sexual difference does, too ('there was scarcely anything left of body or mind by which one could say "This is he" or "This is she"', p. 144). The shawl too begins to unwind: 'once in the middle of the night with a roar, with a rupture . . . one fold of the shawl loosened and swung to and fro' (p. 148); 'as the long stroke lay upon the bed . . . another fold of the shawl loosened' (p. 151). 'Whatever they hung that beast's skull there for?' (p. 160), asks the indomitable old cleaning lady, herself a scarcely human figure of abjection who carries on, parodically, Mrs Ramsay's homemaking function in her absence. The unswathed skull becomes an emblem of what Mrs Ramsay, and Woolf with her, have simultaneously veiled and displayed with their web of words – the process that separates the not-yet-subject from its not-yet-object, the (ever-absent) mother. In Kristeva's account, it is this process that makes signification possible by way of the emerging subject's identification with the signs which stand in the mother's empty place. *To the Lighthouse* inscribes the movement of abjection without which there could be no subjectivity, and no signification either. Differentiation begins here; within the representational scheme of the novel, painting figuratively occupies the position of the sign-system – language – which the text actually employs. If Lily's line at the end of the novel is the emblem of minimal but fixed difference which secures her self-inscription, the price Lily pays for finishing her picture is the casting out of the mother, her beloved Mrs Ramsay.[42] Or Mrs Ramsay dies suddenly so that the 'third stroke' may be appropriated not only for Lily's art, but for Woolf's writing. The pre-Oedipal configuration which (re)produces the mother as the origin of all signification in Kristevan theory not only allows Woolf's novel to read – to be read – beyond Freudian theory, but suggests how a reading of Woolf might revise and extend feminist thinking about the pre-Oedipal. Such a reading of Mrs Ramsay's 'third stroke' (like the lighthouse, 'so much her, yet so little her') points, however, to the crucial internal difference between the feminist reading and the feminist theory that makes it possible – or between a fiction and the theory it sustains. Like screen memory, the theoretical activity which I've called 'Reading Woolf with Freud' is inevitably 'tendentious', anything but indifferent in its attempt to rescue *To the Lighthouse* for an alternative feminist reading of both Woolf and Freud – a reading that emphasizes the role of the pre-Oedipal in order to offer an account of the relations between sexual difference, subjectivity, and writing.

Notes

1. *The Standard Edition of the Complete Psychological Works of Sigmund Freud*, ed. James Strachey, 24 vols (London: Hogarth Press, 1953–74), 17.245; cited hereafter as *SE*, followed by volume and page nos.

2. *A Writer's Diary*, ed. Leonard Woolf (London: Hogarth Press, 1953), p. 76.

3. *Sexual/Textual Politics: Feminist Literary Theory* (London and New York: Methuen, 1985), pp. 9–10.

4. *To the Lighthouse* (Harmondsworth: Penguin, 1964), p. 99; subsequent parenthetical page numbers in the text refer to this edition.

5. See *Powers of Horror: An Essay on Abjection*, trans. Leon S. Roudiez (New York: Columbia University Press, 1982); and ':' L'abject d'amour', translated as 'Freud and Love: Treatment and its Discontents', in *The Kristeva Reader*, ed. Toril Moi (New York: Columbia University, 1986), pp. 238–71.

6. See 'Unmaking and Making in *To the Lighthouse*' in Sally McConell–Ginet, Ruth Borker and Nelly Furman (eds), *Women and Language in Literature and Society* (New York: Praeger, 1980), p. 313.

7. See 'Some Psychical Consequences of the Anatomical Distinction Between the Sexes' (1925), *SE*, 19.243–58.

8. See 'Fetishism' (1927), *SE*, 21.149–57.

9. *The Psychopathology of Everyday Life* (1901) *SE*, 6.48.

10. See *Feminine Sexuality: Jacques Lacan and the école freudienne*, ed. Juliet Mitchell and Jacqueline Rose (New York: Norton, 1982), pp. 74–85.

11. 'Screen Memories' (1989), *SE*, 3.311.

12. *SE*, 3.313.

13. *SE*, 3.314.

14. *SE*, 3.316.

15. *SE*, 3.317.

16. *SE*, 3.318.

17. *SE*, 3.318.

18. *The Complete Letters of Sigmund Freud to Wilhelm Fliess 1887–1904*, trans. and ed. Jeffrey Moussaieff Masson (Cambridge, Mass.: Harvard University Press, 1985).

19. Ibid., p. 249.

20. Ibid., p. 249.

21. Ovid, *Metamorphoses*, Book V, trans. Frank Justus Miller, Loeb edn, 2 vols (London, 1916), 2. 265–7.

22. See, for instance, 'Katharina', *Studies in Hysteria* (1895), *SE*, 2.134 *n.*, where Freud later admitted that 'uncle' had screened an incestuous attempt at seduction by the father.

23. *Of Woman Born: Motherhood as Experience and Institution* (New York: Norton. 1976), p. 226.

24. *Women and Madness* (New York: Avon, 1976), p. 17.

25. 'Beyond Oedipus: The Specimen Story of Psychoanalysis', *Modern Language Notes*, **98**, (1983), 1046.

26. *Reading Lacan* (Ithaca: Cornell, University Press and London 1985), p. 146 (my italics).

27. Ibid., p. 147.

28. Ibid., pp. 148. 149.

29. *Of Women Born*, p. 228.

30. Ibid., p. 229.

31. Ibid., p. 240.

32. See especially *Powers of Horror*, pp. 1–16, 32–55 *passim*, and 'Freud and Love' ('L'abject d'amour'), *The Kristeva Reader*, pp. 240–8, 256–60.

33. 'Unmaking and Making in *To the Lighthouse*', *Women and Language in Literature and Society*, p. 316.

34. *The Complete Letters of Sigmund Freud to Wilhelm Fliess*, pp. 268, 269.

35. Ibid., p. 269.

36. See *SE*, 5.664.

37. *Powers of Horror*, p. 5.

38. Ibid., p. 46

39. Ibid., p. 38.

40. Ibid., p. 37.

41. Ibid., p. 37.

42. For an account of 'minimal difference' in the context of a theoretical discussion of Kristevan 'abjection' and the mother, see NEIL HERTZ, *The End of the Line: Essays on Psychoanalysis and the Sublime* (New York: Columbia University Press, 1985), pp. 231–7.

7 Tradition and Revision in Woolf's *Orlando*: Defoe and 'The Jessamy Brides'*

Susan M. Squier

Many feminist critics have been interested in re-examining the canon of men writers, whether to insert some female names into it or to question altogether the principles according to which an apparently hierarchical and exclusive order of merit gets constructed in the first place. There has also been considerable interest in looking at how women writers will have to negotiate their own relations to a literary establishment which appears to them as resolutely masculine, and which actively discourages their participation. Squier's essay takes up such questions in relation to Woolf's own literary development, which she sees as having been dominated by the need for her to free herself from various kinds of paternal influence. Her own father was put out of the way by *To the Lighthouse*, the novel Woolf spoke of herself as a work of personal mourning; *Orlando*, the next novel, often regarded as little more than a playful interlude in Woolf's *œuvre*, represents for Squier her achievement of an equivalent freedom in relation to an equally paternal literary tradition.

On 14 March 1927 Virginia Woolf recorded in her diary the symptoms of an 'extremely mysterious process . . . the conception last night between 12 & one of a new book' (*Diary* III, p. 131).

> I sketched the possibilities which an unattractive woman, penniless, alone, might yet bring into being . . . It struck me, vaguely, that I might write a Defoe narrative for fun. Suddenly between twelve & one I conceived a whole fantasy to be called 'The Jessamy Brides' – why, I wonder . . . No attempt is to be made to realise the character. Sapphism is to be suggested. Satire is to be the main note – satire & wilderness.
>
> (D. III, 131)

* Reprinted from *Women's Studies*, **12**, 2 (1986), 167–78.

That Woolf's sixth novel represents an act of comic tribute to her loving friend Vita Sackville-West has been well documented.[1] However, a return to the diary record of *Orlando*'s conception suggests that the work was more than the playful tribute to Sackville-West or the escape from works of a more 'serious poetic experimental' nature which some critics, following Woolf's lead, have judged it to be.[2] Instead, *Orlando* completes the labor of self-creation which Woolf began in her autobiographical novel, *To the Lighthouse* (1927). In that earlier novel, she laid to rest the ghosts of her parents, establishing herself as an adult woman independent of their potentially eclipsing examples. In *Orlando* (1928), which bears 'the clear stamp of her mind in its maturity', and 'might in a sense have been called an autobiography', Woolf went further.[3] Claiming her *literary* majority, she confronted the influence of both literal and literary fathers to reshape the novel, and so to create a place for herself in the English novelistic tradition which was their legacy to her. *Orlando* may be read as a serious work of criticism as well as a love-tribute, then, and its major concerns are prefigured in Woolf's diary entry of 14 March 1927: the confrontation with one's literary precursors and the impact of gender upon literary genre.[4]

With *Orlando*'s subtitle, 'A Biography', Woolf issued a challenge both to her father, Sir Leslie Stephen – who had achieved his prominence as the editor of the *Dictionary of National Biography* – and more broadly to the patrilineal tradition of English Literature which Stephen traced in his important volume, *English Literature and Society in the Eighteenth Century*.[5] Preeminent in that tradition was Daniel Defoe, who to Leslie Stephen was the father of the English novel. Stephen wrote of Defoe as typifying the middle-class values which led to the rise of the novel as genre; individualism, nationalism, and resourceful pragmatism.[6] But Defoe was also the aesthetic father of *Orlando*. To Woolf, Defoe led the list – as she announced in the Preface to *Orlando* – of those 'dead and so illustrious that I scarcely dare name them, yet no one can read or write without being perpetually in their debt' (*Orlando*, p. 5). Woolf's homage to Defoe, like her relationship with Leslie Stephen, was far from simple, however; consideration of Defoe's role as acknowledged precursor of *Orlando* will reveal the scope and nature of Woolf's designs on the English novel. In acknowledging her debt to Defoe, Woolf was also subverting his influence and challenging the genre of the realistic novel which he initiated. Moreover, by challenging that genre, she framed a gender-based critique of the patriarchal ideology in which the novel has its origins.

In her bicentenary essay on Daniel Defoe, Woolf wrote of him as 'one of the first indeed to shape the novel and launch it on its way' (*Collected Essays*, I, p. 90) This position of precedence makes itself felt in *Orlando*, which in both narration and plot recalls Defoe's *Moll Flanders*. As early as

1919 Woolf had praised that novel for the 'briskness of the story', and her own *Orlando* – written less than a decade later – shared the light picaresque mode and brisk pacing of Defoe's novel, as well as its episodic plot (CE, I, p. 92). Aspects of the characters also resemble each other: both novels have protagonists who disguise themselves as men; who consort with prostitutes and gipsies; who are experienced and capable international travellers; who are mothers (Moll, repeatedly); and who explore the different strata of London society. Yet their stories seem to be inverse mirror images: Moll Flanders develops from a position of social marginality (as a pickpocket, whore, and bigamist) to a penitent identification with conventional social values, while Orlando moves from a position of privileged centrality (due to his aristocratic lineage, great hereditary wealth, and masculine prerogative) to a position of social marginality as a woman and poet, a position both character and novel affirm. Despite the characters' different developmental paths, Woolf's comments in 'Defoe' suggests that she borrowed the spirit of *Orlando* from *Moll Flanders*, for she asserts there that Defoe's work contained a subterranean affirmation of his protagonist which the author's overt moralizing belied.

> The interpretation that we put on his characters might . . . well have puzzled him. We find for ourselves meanings which he was careful to disguise even from himself. Thus it comes about that we admire Moll Flanders far more than we blame her. Nor can we believe that Defoe had made up his mind to the precise degree of her guilt, or was unaware that in considering the lives of the abandoned he raised many deep questions and hinted, if he did not state, answers quite at variance with his professions of belief.
>
> (CE, I, p. 94)

In choosing a 'Defoe narrative' as her model for *Orlando*, Woolf was engaged in the same technique of revisionary reading that she demonstrated in 'Defoe'. She chose to find in Defoe's novel meanings which, while they might have puzzled him, were nonetheless crucial to her at that stage of her artistic development. Admiring Moll rather than blaming her, Woolf found in the character qualities upon which she modeled her own literary emancipation, for that was the task at hand following her completion of *To the Lighthouse*.

As Woolf understood *Moll Flanders*, the novel's admirable spirit of briskness was due 'partly to the fact that having transgressed the accepted laws at a very early age she [Moll] has henceforth the freedom of the outcast' (CE, I, p. 92). The idea that alienation could liberate was appealing to Woolf, who from her early years had struggled with feelings of exclusion from social and literary London.[7] By the time she wrote *Three*

Guineas (1938) she firmly believed that the outsider position was the best vantage point from which to work for peace and sexual equality. However, roughly ten years earlier her mind was not yet made up on this issue, and she found in *Orlando* an opportunity to try out the outsider's experience – not only in the person of her protagonist, but also in the novel's very form and style. Like *Moll Flanders*, *Orlando* transgresses 'accepted laws', but those laws are not only social, but literary.

Consideration of the rules of characterization which *Orlando* contravened demonstrates the novel's liberating revision of the genre 'launched' by Defoe. A crucial challenge to writers in the novel's early years was the creation of individuated characters, rather than the stock characters, allegorical figures, or types which peopled earlier literary genres. Four techniques contributed to the creation of this new character, according to Ian Watt: a detailed presentation of characters in relation to their environment; the use of past experience as cause of present action (in contrast to the earlier stress on disguise and coincidence in plotting); a more minutely discriminated time scale; and finally the use of realistic proper (first and last) names, rather than emblematic or conventional single names.[8] Woolf's earliest plans for *Orlando* expressed her opposition to the cardinal tenet of the developing novel tradition, the creation of individuated characters: 'No attempt', she specified in her diary entry of March 1927, 'is to be made to realise the character.' (D, III, p. 131) Moreover, throughout *Orlando* she maintained her opposition to traditional novelistic practice, by inverting all of the techniques for character creation in order not to emphasize identity, but to call the very concept into question. So, Orlando's single name (hereditary titles aside) makes him/her the very type of comic gender reversal, in its allusion to Shakespeare's *As You Like It*.[9] The more minutely discriminated time scale of the realistic novel here becomes a life of more than five hundred years which alternately flash by without commentary and creep along second by second. While in the realistic novel the plot introduced a new logic of cause and effect; in *Orlando* the preeminent dramatic action – Orlando's gender change – seems uncaused by any previous event, and is accompanied both by coincidence (the trance occurs at the same time as the Turkish revolt against the Sultan) and disguise (Orlando passes for a Turkish boy and a gipsy after the gender change). Finally, the realistic novel's careful presentation of character in relation to environment becomes, in *Orlando*, a survey of the protagonist's contrasting experiences of Elizabethan, Jacobean, Carolinean, Restoration, Augustan, Victorian and Modern England, as well as of Constantinople before and after the Sultan's fall. This extensive topographical and temporal detail, far from increasing our belief in Orlando as an actual person, rather causes us to view him/her as the type or symbol of the British poet, nobleman/woman and statesman.

Woolf's treatment of Orlando's character thus demonstrates her divergence from what she has seen, in 1919, as the key ingredient of the Defoe narrative – the conviction that 'the novel had to justify its existence by telling a true story', or what Ian Watt has defined as the prnciple of 'formal realism':

> the premise, or primary convention, that the novel is a full and authentic report of human experience and is therefore under an obligation to satisfy its reader with such details of the story as the individuality of the actors concerned, the particulars of the times and places of their actions, details which are presented through a more largely referential use of language than is common in other literary forms.[10]

Intimations of Woolf's motives for so diverging from Defoe's novelistic technique appear in her 1919 essay. There, she wrote of Defoe's works as 'prosaic', characteristic of the 'great plain writers, whose work is founded upon a knowledge of what is most persistent, though not most seductive, in human nature.' (CE, i, p. 97) Her view recalls Leslie Stephen's indictment of Defoe for 'absence of any passion or sentiment', and his dry judgement that 'the merit of De Foe's narrative bears a direct proportion to the intrinsic merit of a plain statement of the facts'.[11] Yet a more recent and deeply felt influence than Leslie Stephen's prompted Woolf to subvert Defoe's realism and plain speaking. The sexual metaphor buried in the prose of both father and daughter – the notion of Defoe as lacking a certain sexual quality, as being neither seductive, passionate, nor full of sentiment – suggests that sexuality or gender may have played a part in Woolf's decision to revise Defoe's style. Indeed, a connection may be made between Defoe's prose style and character creation and the world view of the literary tradition Defoe fathered, suggesting that it was a desire to reevaluate the patriarchal literary tradition and culture – under the influence of her new love for Vita Sackville-West – which promoted the form and style of *Orlando*.

When Vita first read *Orlando*, in October 1928, she wrote to Virginia Woolf, 'I am completely dazzled, bewitched, enchanted, under a spell' (*Letters*, iii, p.573). The novel's style brought to her mind 'a robe stitched with jewels' (*Letters*, iii, p. 573). Sackville-West's comments indicate the elements in *Orlando* which most fully subvert the influence of Defoe: *Orlando*'s unrealistic, even fantastical and magical plot and character, and *Orlando*'s ornate diction. Moreover, Vita's comments suggest Woolf's reason for challenging the patriarchal bias of the novel genre in *Orlando*:

her novel's celebration of qualities unacceptable to traditional realism. The eighteenth-century fathers of the novel adhered to a strictly referential use of language, corollary to the commitment to formal realism which aligned them with the realist philosophers.[12] That this linguistic probity had moral and cultural ramifications for them was most tellingly expressed in Locke's observation, in the *Essay Concerning Human Understanding*, that 'eloquence, like the fair sex, involves a pleasurable deceit'.[13] This assumed connection between eloquence, deceit and femininity which lay at the basis of formal realism would have made such a technique inappropriate for *Orlando*, a novel written to celebrate the seductive, passionate character of Vita Sackville-West, by a writer who admitted, 'It is true that I only want to show off to women. Women alone stir my imagination' (L, IV, p. 203). For if Defoe's realism eschewed elaborate diction as deceitful, seductive, and above all feminine, Woolf turned in *Orlando* to elaborate diction and fanciful plotting precisely because it could best embody the complex woman poet to whom the novel paid homage. Woolf turned the realistic novel on its head in *Orlando*, playing with distinctions between reality and fantasy, truth and lying, masculinity and femininity. Her challenge to the conventions of realism, boldy emphasized by her prefatory tribute to Defoe, reflects the realization that the rules of genre are shaped by the politics of gender. As Nancy Miller has pointed out,

> The attack on female plots and plausibilities assumes that women writers cannot or will not obey the rules of fiction. It also assumes that the truth devolving from *veri*similitude is male. For sensibility, sensitivity, 'extravagance' – so many code words for feminine in our culture that the attack is in fact tautological – are taken to be not merely inferior modalities of production but deviations from some obvious truth. The blind spot here is both political (or philosophical) and literary. It does not see, nor does it want to, that the fictions of desire behind the desiderata of fiction are masculine and not universal constructs.[14]

Challenging the universality of the 'fictions of desire' basic to the realistic novel, Woolf was both following and subverting her literary and literal fathers. Not only did she contravene Defoe's style of moralistic truth-telling nearly point-for-point, but she also revised Stephen's approach to literary criticism, in which he had been – according to Noel Annan – the first writer to consider the impact of changing social classes upon the production of literature and the development of new literary genres.[15] By shifting her index from his focus on economic and social class to her own interest in sex class, Woolf subverted her father's approach in order to consider specifically the impact of *gender* upon *genre*. While all of *Orlando* is a meditation upon this issue, perhaps the most pointed statement of Woolf's concern comes near the novel's end, in one of the 'biographer's' numerous asides to the reader (and we might mark Woolf's sly glee at putting such a speech in the mouth of one of her father's fellow biographers):

> we must here snatch time to remark how discomposing it is for the biographers that this culmination and peroration should be dashed from us on a laugh casually like this; but the truth is that when we write of a woman, everything is out of place – culminations and perorations; the accent never falls where it does with a man.
>
> (*Orlando*, p. 204)

With the veiled allusion to other sorts of climaxes, Woolf intimates the sexual and gender-related motives for her revision of her father's art as biographer and literary critic. To the novelist/biographer writing of a woman, 'truth' takes a different form than it does with a man. Woolf's metaphor encompasses not only the multiple climaxes of the female orgasm, but the multiple high-points of Orlando's life – the birth of her child, the publication of 'The Oak Tree', her marriage, winning the Burdett Coutts' Memorial Prize for her poem. Furthermore, Woolf's metaphor suggests that language, like the shape of a literary work, shifts when it is called upon to describe not the man around whom a novel or biography is typically composed, but rather the woman who discomposes it. Working in the *spirit* of her father's criticism, Woolf subverted the *matter*, calling into question both gender divisions and the literary form enshrining them.

The sexually liberating nature of *Orlando's* homage to Defoe – and the realistic novel he fathered – lies in the fact that to challenge the constraints of the literary canon is to confront the deepest politics or philosophy linking gender to genre. As Nancy Miller has pointed out, both style and character creation reflect ideology:

> To build a narrative around a character whose behaviour is deliberately idiopathic . . . is not merely to create a puzzling fiction, but to fly in the face of a certain ideology (of the text and its context), to violate a grammar of motives that describes while prescribing . . . what wives, not to say women, should or should not do.[16]

In beginning with the idea 'an unattractive woman, penniless, alone', Woolf in her choice of a heroine defied the ideology ascribing to a woman value depending upon her worth to a man (D, III, p. 131). It is this ideology which Mrs Ramsay espouses, in *To the Lighthouse*, when she

127

urges 'Minta must, they all must marry, since . . . an unmarried woman has missed the best of life' (*To the Lighthouse*, p. 77). And by settling on Orlando, a character whose gender changes midway through the novel, Woolf further violates the ideology shaping patriarchal culture itself: what Gayle Rubin has termed the 'sex/gender system'.[17] Woolf both invoked and defied this system when she first planned to title the novel 'The Jessamy Brides,' (D, III, p. 131). 'Jessamy', 'a man who scents himself with perfume or who wears a sprig of Jessamine in his buttonhole . . . a dandy, a fop', had by the time Woolf wrote *Orlando* been paired conceptually with an equally transgressive parallel term, 'Amazon', a woman whose stature and physical prowess were more conventionally masculine than feminine.[18] In its evocation of the gender line crossing embodied by a foppish male bride, a feminine man, and an 'Amazonian' woman, the initial title for Orlando prefigures Woolf's protagonist, who in the course of a long life loved both sexes passionately, contracted a marriage to a man whom she jokingly suspected of being a woman (and who entertained the corresponding suspicion that Orlando was a man), and who at the novel's conclusion summons herself ('Orlando?') only to be answered by a multiplicity of selves – of both genders and sexual orientations (*Orlando*, pp. 201–2) The provisional title also brings to mind the lovingly unorthodox marriage of Vita Sackville-West and Harold Nicolson, which encompassed homosexual affairs for both parties while remaining their primary and most honored tie. In its allusions to a variety of unorthodox marriages, 'The Jessamy Brides' not only embodies the disobedient briskness which Woolf admired in *Moll Flanders*, but it replicates Woolf's own defiance of social conventions in her love affair with Vita Sackville-West.

That love affair was already underway when Woolf finished her autobiographical novel, *To the Lighthouse*, and she gave Vita an elaborately bound copy of the novel inscribed, 'Vita from Virginia (In my opinion the best novel I have ever written.)' Yet when Vita opened the expensively bound volume, she found only blank pages. This gesture of the blank-paged book both encompasses the past and anticipates the future, supporting what Louise DeSalvo has cogently documented: that the relationship with Vita was for Woolf an important artistic turning-point. Underscoring the homage to Woolf's two mothers, Julia Stephen and now Vita Sackville-West, in its evocative blank pages this gift also intimates the coming challenge to two fathers, Leslie Stephen and Daniel Defoe, in the rupture with the patriarchal novelistic tradition to come. For the gift of this blank-paged book suggests that whatever Woolf's 'best novel' is to be, if it is to reflect her love for Vita Sackville-West it will be unreadable within the patriarchal tradition of English literature and culture. And indeed, even the conception of that novel still to come challenged the form at the heart of *To the Lighthouse*, 'father and mother and child in the garden' (*A Writer's Diary*, 79). For having found herself 'virgin, passive, blank of ideas' for a number of weeks following the completion of *To the Lighthouse*. Woolf on that mysterious March night of 1927 found stirring within her *Orlando* – the consummation of her love for Vita. In its defiance of the 'Defoe narrative' as in its evocative embodiment of another form for both gender and genre, *Orlando* stands not only as a serious work of criticism *in and of* the tradition of Leslie Stephen, but as Woolf's gesture of literary emancipation.

Notes

1. See MADELINE MOORE, '*Orlando*: An Edition of the Manuscript', *Twentieth-Century Literature*, **25**, 3/4 (Fall/Winter 1979)., 303–55, especially appendices B and C; LOUISE A. DESALVO, 'Lighting the Cave: The Relationship between Vita Sackville-West and Virginia Woolf', *Signs*, **8**, 2 (Winter, 1982), 195–214; JEAN O. LOVE, '*Orlando* and its Genesis: Venturing and Experimenting in Art, Love, and Sex', in *Virginia Woolf: Revaluation and Continuity*, ed. Ralph Freedman (Berkeley and Los Angeles: University of California Press, 1980); J. J. WILSON, 'Why is *Orlando* Difficult?' *New Feminist Essays on Virginia Woolf*, ed. Jane Marcus (London: The Macmillan Press, 1981), 170–84.

2. Writing in her diary, Woolf in March 1927 welcomed *Orlando*, acknowledging that 'the truth is I feel the need of an escapade after all these serious poetic experimental books whose form is always so closely considered'. In October of the same year she wrote, 'Talk of planning a book, or waiting for an idea! This one came in a rush; I said to pacify myself, being bored & stale with criticism . . . "You shall write a page of a story for a treat: you shall stop sharp at 11.30 & then go on with the Romantics." I had very little idea what the story was to be about. But the relief of turning my mind that way about was such that I felt happier than for months . . .' *The Diary of Virginia Woolf*, Vol. III, ed. Anne Olivier Bell (New York: Harcourt Brace Jovanovich, 1980), pp. 131, 161–2. For contemporary critical opinions of *Orlando* following its publication, see *Virginia Woolf: The Critical Heritage*, ed. Robin Majumdar and Allen McLaurin (London: Routledge & Kegan Paul, 1975)

3. HELEN MACAFEE, initialled review, *Yale Review*, **18**, xvi (1929), reprinted in *Virginia Woolf: The Critical Heritage*, p. 237.

4. This is not to claim that Woolf consciously intended such a serious focus, but rather that the mysterious serendipity of *Orlando*'s conception veiled from her a program which represented a continuation – even a culmination – of Woolf's concerns in *To the Lighthouse*: a woman artist's struggle for social and aesthetic emancipation from the ideology of the Victorian patriarchal family.

5. *Orlando* abounds in such challenges to the biographer, from its opening sentence, 'He – for there could be no doubt of his sex . . .' with its hint of the antics to come, to Woolf's jab – near the end of the novel – at her father's particular occupation, as compiler of the *Dictionary of National Biography*: 'The true length of a person's life, whatever the *Dictionary of National Biography* may say, is always a matter of dispute' (*Orlando*, pp. 7, 200). In fact, the very

form of the novel – one might call it a biography inside out – subverts the biographer's authority, constantly asserting the unknowability of major facts and the precision of minor ones. For discussion of Stephen's influence on Woolf's intellectual life, see KATHERINE C. HILL, 'Virginia Woolf and Leslie Stephen: History and Literary Revolution', *PMLA*, **96**, 3 (May, 1981), 351–62. References to the essay of 1919 on Defoe are taken from the *Collected Essays*, ed. Leonard Woolf, vol. I (London: Chatto and Windus, 1966).

6. STEPHEN, *English Literature and Society in the Eighteenth Century* (London: Gerald Duckworth, 1904), pp. 135–6. 'Defoe, as the most thorough type of the English class to which he belonged, could not do otherwise than to make his creation a perfect embodiment of his own qualities. *Robinson Crusoe* . . . has supplied innumerable illustrations to writers on Political Economy. One reason is that Crusoe is the very incarnation of individualism . . . This exemplary person not only embodies the type of middle-class Briton but represents his most romantic aspirations . . .' See also HILL, pp. 356–7.

7. Woolf's memoir, 'A Sketch of the Past', contains one of the classic formulations of this feeling, as she experienced it in her relationship with her half-brother, George Duckworth: 'He was thirty-six when I was twenty. He had a thousand pounds a year and I had fifty . . .But besides feeling his age and his power, I felt too another feeling which I later called the outsider's feeling. When exposed to George's scowling, I felt as a tramp or a gipsy must feel who stands at the flap of a tent and sees the circus going on inside. Victorian society was in full swing; George was the acrobat who jumped through hoops, and Vanessa and I beheld the spectacle. We had good seats at the show, but we were not allowed to take part in it. We applauded, we obeyed – that was all' (*Moments of Being*, ed. Jeanne Schulkind (New York: Harcourt Brace Jovanovich, 1976), pp. 131–2). For a further discussion of this passage in relation to Woolf's treatment of this outsider feeling in her fiction, see SUSAN SQUIER, *Virginia Woolf and London: The Sexual Politics of the City* (Chapel Hill: The University of North Carolina Press, 1985).

8. IAN WATT, *The Rise of the Novel: Studies in Defoe, Richardson and Fielding* (Berkeley: The University of California Press, 1971) pp. 18–27.

9. BEVERLY Ann SCHLACK, *Continuing Presences: Virginia Woolf's Use of Literary Allusion* (University Park: The Pennsylvania State University Press, 1979), p. 80.

10. WOOLF, *The Common Reader* (New York: Harcourt, Brace & World, Inc., 1953), pp. 90–1; WATT, p. 32.

11. SIR LESLIE STEPHEN, 'Defoe's Novels', *Moll Flanders*, ed. Edward Kelly (New York: W. W. Norton & Co., Inc., 1973), p. 333.

12. WATT, pp. 27–32.

13. Ibid., p. 28.

14. NANCY K. MILLER, 'Emphasis Added: Plots and Plausibilities in Women's Fiction', *PMLA*, **96**, 1 (January, 1981), 36–48, 46.

15. NOEL ANNAN, *Leslie Stephen: His Thoughts and Character in Relation to His Time* (Cambridge: Harvard University Press, 1952), 270. See also Katherine C. Hill, pp. 356–7.

16. MILLER, p. 37.

17. 'The Traffic in Women', *Towards an Anthropology of Women*, ed. Rayna Reiter (New York: Monthly Review Press, 1975), pp. 157–210; reprinted in *Feminist Frameworks*, ed. Alison M. Jaggar and Paula Rothenberg Struhl (New York: McGraw-Hill Book Co., 1978), pp. 154–67, 155. 'As a preliminary definition, a "sex/gender system" is the set of arrangements by which a society transforms biological sexuality into products of human activity, and in which these transformed sexual needs are satisfied.'

18. *The Compact Edition of the Oxford English Dictionary*, Vol. I (Oxford: Oxford University Press, 1971), p. 1506.

8 The Island and the Aeroplane: The Case of Virginia Woolf*

GILLIAN BEER

This piece shows the opportunities for bringing historical criticism
and feminist criticism together around a topic which – so Beer's piece
suggests – will be seen in many ways differently from a woman's
perspective. The island is bound up with a whole tradition of thinking
about England and English identity which was set to suffer encroach-
ments from that modern machine unbound to either natural or
fictional shorelines. The image of the aeroplane, which at the time
was at the height of its power to inspire collective fantasies, is utilised
by Woolf as part of her idiosyncratic feminist critique of insular
masculine Englishness and militaristic aggressiveness, as well as in
relation to a rather different set of associations with feminine versions
of ambitious or erotic soaring. At the same time, the break-up of
clear-cut boundaries of nation, body, self, as well as of narrative and
syntactical order, which Woolf's writing enacts, has sharp links to a
more fluid politics figured and enabled by the mutations of perspective implied by the loss of the island myth and the rise of the plane.

England's is, so writers over the centuries have assured us, an island
story. What happened to that story with the coming of the aeroplane?
That larger question is central to my essay but my chosen example is a
particular one: the writing of Virginia Woolf.

The advent of the aeroplane had profound political and economic
consequences;[1] the object itself rapidly entered the repertoire of dream
symbols, with their capacity for expressing erotic politics and desires. In
the period between 1900 and 1916 Freud came to recognize the extent to
which 'balloons, flying-machines and most recently Zeppelin airships'
had been incorporated into two kinds of dream symbol: those which are
'constructed by an individual out of his own ideational material' and
those 'whose relation to sexual ideas appears to reach back into the very
earliest ages and to the most obscure depths of our conceptual

* Reprinted from Homi K. Bhabha (ed.), *Nation and Narration* (London:
Routledge, 1990), pp. 255–90.

functioning'.[2] His analysis of flying dreams is phallocentric (women can have them as the 'fulfilment of the wish to be a man', a wish which can be realized by means of the clitoris which provides 'the same sensations as men'). He identifies 'the remarkable characteristic of the male organ which enables it to rise up in defiance of the laws of gravity' as the reason for its symbolic representation as a flying-machine. If we pursue that line of argument what are we to make of Virginia Woolf's striking interest in her *Diary* in air crashes? Or of her description of the Zeppelin with an umbilical 'string of light hanging from its navel'?[3] Suffice it for the moment to say that the aeroplane in Woolf's novels is given a crucial presence in four of her works, *Mrs Dalloway, Orlando, The Years* and *Between the Acts*, all of which are concerned with the representation of England and with difficult moments of historical or national change.

The destructiveness and the new beauty generated by the possibilities of flight are realized by Gertrude Stein in her book, *Picasso* (1938), in which she comments on the formal reordering of the earth when seen from the aeroplane – a reordering which does away with centrality and very largely with borders. It is an ordering at the opposite extreme from that of the island, in which centrality is emphasized and the enclosure of land within surrounding shores is the controlling meaning. Stein writes of the First World War thus:

> Really the composition of this war, 1914–18, was not the composition of all previous wars, the composition was not a composition in which there was one man in the center surrounded by a lot of other men but a composition that had neither a beginning nor an end, a composition of which one corner was as important as another corner, in fact the composition of cubism.

Flying over America she thinks: 'the twentieth century is a century which sees the earth as no one has ever seen it, the earth has a splendor that it never has had, and as everything destroys itself in the twentieth century and nothing continues, so then the twentieth century has a splendor, which is its own'.[4] The patchwork continuity of an earth seen in this style undermines the concept of nationhood which relies upon the cultural idea of the island – and undermines, too, the notion of the book as an island. Narrative is no longer held within the determining contours of land-space. Woolf's first novel is *The Voyage Out*, which opens with the ship leaving England – a journey from which its heroine never returns. *Between the Acts*, her last novel, takes up the multiple signification of the island, including that of the literary canon, and places them under the scrutiny of aeroplanes at the beginning and end of the work.

Woolf's quarrel with patriarchy and imperialism gave a particular complexity to her appropriations of the island story. At the same time her symbolizing imagination played upon its multiple significations – land and water margins, home, body, individualism, literary canon – and set them in shifting relations to air and aeroplane.

'If one spirit animates the whole, what about the aeroplanes?' queries a character in Woolf's last novel, *Between the Acts*, which is set on a day in mid-June 1939 and was written with the Battle of Britain going on overhead.[5] In the midst of its composition a last version of the island story as safe fortress was played out: an 'armada' (telling word) of little boats set out from England to rescue from the beaches of Dunkirk the British soldiers being strafed by German bombers. Woolf writes of the events with great intensity in her diary, as we shall see. In the public mythologization of that episode it has never been quite clear whether the topic is triumph or defeat.

Much earlier in the century H. G. Wells, concluded his novel of coming events, *The War in the Air*, with Bert Smallways thinking that 'the little island in the silver sea was at the end of its immunity'.[6] Yet the myth of the fortress-island was sustained past the beginning of the Second World War, so that the sculptor and refugee Naum Gabo could note in his Diary in 1941:

The sea lies stark naked between my windows and the horizon . . .
The heart suffers looking at it and the contrast with what is happening in the world . . . how many more weeks will this peace last on this little plot of land? . . . Our life on this island, in this last fortress of the old Europe, gradually enters . . . into a state of siege.[7]

In *British Aviation: The Ominous Skies* 1935–39 Harald Penrose reports a journalist as writing at the beginning of August 1939, a month before the outbreak of the Second World War: 'The dangers of air attack have been much magnified. This country is protected by stretches of sea too wide for the enemy to have an effective escort of fighters.'[8]

The advent of the aeroplane was by no means only a military phenomenon, of course. H. G. Wells was not far out when in the 1890s he wrote a forward fantasy, 'Filmer', in which the hero's mastery of the art of flying 'pressed the button that has changed peace and warfare and well nigh every condition of human life and happiness'.[9] In 'The argonauts of the air' he grimly foresaw the future of the flying machine: 'In lives and in treasure the cost of the conquest of the empire of the air may even exceed all that has been spent in man's great conquest of the sea.'[10] Later in his life, when he had experienced the pleasures of flying for himself, and when the diversity of uses for the aeroplane had become actual, Wells argues against the individualism of light aircraft and for large passenger airships in order that the freedoms and pleasures of flight should be opened to many. In 'The present uselessness and danger of aeroplanes. A problem in organization', in *The Way the World is Going* (1928), he writes:

> I know the happiness and wonder of flying, and I know that its present rarity, danger, and unattractiveness are not due to any defects in the aeroplane or airship itself – physical science and mechanical invention have failed at no point in the matter – but mainly, almost entirely, to the financial, administrative, and political difficulties of aviation.[11]

Unlike H. G. Wells and Gertrude Stein, Virginia Woolf never flew, though she fantasized the experience vividly in her late essay 'Flying over London'.[12] But the aeroplane is powerfully placed in her novels. It typifies the present day, and beyond that it is a bearer and breaker of signification, puffing dissolving words into the air in *Mrs Dalloway* to be diversely construed by all the casual watchers of its commercial task. In *Between the Acts* its presence is more menacing, breaking apart the synthesizing words of the rector at the end of the village pageant, rumbling overhead towards the imminent war. *The Years* represents the whole period of the First World War by the 1917 air raid. ('The first mass aeroplane raid took place on London on June 13, 1917', Gibbs-Smith informs us.)[13] *Orlando* ends with the sea-captain-husband transformed into an aeronaut, 'hovering' over Orlando's head. Menace, community, eroticism, warfare and idle beauty: the aeroplane moves freely across all these zones in her writing. The pilot's eye offers a new position for narrative distance which resolves (as at the opening of *The Years*) the scanned plurality of the community below into patterns and repeats. Woolf was not alone in her invocation of the aeroplane, of course: we think across immediately to Yeats and Auden. But she was, I think, particularly acute in her understanding of it in relation to the cultural form of the island, and extraordinarily economical in her appraisal.

The story of Daedalus and Icarus – the craftsman father who made the flying machine and the flying son whose wings loosened disastrously when he flew too near the sun, which melted the wax that attached them – revives in twentieth-century literature, very possibly accompanying the coming of aeroplanes. In the last pages of *Portrait of the Artist*, written in 1914, such imagery seems still entirely mythological and is related to ships sailing rather than aircraft flying.

April 16. Away! Away!
The spell of arms and voices: the white arms of roads, their promise of close embraces and the black arms of tall ships that stand against the moon, their tale of distant nations. They are held out to say: We are alone – come. And the voices say with them: We are your kinsmen.

And the air is thick with their company as they call to me, their
kinsman, making ready to go, shaking the wings of their exultant and
terrible youth.[14]

In 'Musée des Beaux Arts' (1938) Auden makes of Breughel's *Icarus* an
image of the insouciance with which suffering is surrounded, 'its human
position'.

In Breughel's *Icarus*, for instance; how everything turns away
Quite leisurely from the disaster; the ploughman may
Have heard the splash, the forsaken cry,
But for him it was not an important failure; the sun shone
As it had to on the white legs disappearing into the green
Water; and the expensive delicate ship that must have seen
Something amazing, a boy falling out of the sky,
Had somewhere to get to and sailed calmly on.

(*December 1938*)[15]

The discreet clarity of description here distances the disaster and mutes
its specific reference to Europe at the time of its inscribed date 'December
1938', keeping the discrepancy between acute suffering and humdrum
life permanently disturbing. In another poem written in the same years
(which Auden did not retain in 'Sonnets from China' but which appears
in the original sequence, *In Time of War*, as sonnet 15) Auden
concentrates on the pilots 'remote like savants' who are preoccupied only
with skill as they approach the city. (Fuller reads this poem as being
about politicians, surely mistakenly.)

Engines bear them through the sky: they're free
And isolated like the very rich;
Remote like savants, they can only see
The breathing city as a target which

Requires their skill; will never see how flying
Is the creation of ideas they hate,
Nor how their own machines are always trying
To push through into life. They chose a fate

The islands where they live did not compel.[16]

Auden and Yeats dwell on the human pilots of such planes: the isolation
and aggression of such a position is inimical to Woolf, for it is
dangerously caught in to the militarism it castigates. But she, like them,
does explore the paradoxes in flight:

how their own machines are always trying
To push through into life (as Auden writes).

The glamour that pervades 'An Irish Airman foresees his Death' in *The Wild Swans at Coole* (1919) is riposted even within Yeats' own work in a poem called 'Reprisals', not printed in the definitive edition and first published after the Second World War, in 1948. It opens:

Some nineteen German planes, they say,
You had brought down before you died.
We called it a good death. Today
Can ghost or man be satisfied?[17]

One of Woolf's last essays, written as she was finishing *Between the Acts*, was 'Thoughts on peace in an air raid', prepared for an American symposium 'on current matters concerning women'.[18]

Woolf's opposition to patriarchy and imperialism, her determined assertion that she was 'no patriot', her emphasis on women's 'difference of view', all have their bearing on her figuring of the aeroplane and of flight in her writing. The diaries and the essays, as well as the novels, allow us, through her, to understand some of the ways in which the advent of the aeroplane reordered the axes of experience.

The island

The identification of England with the island is already, and from the start, a fiction. It is a fiction, but an unwavering one among English writers and other English people, that England occupies the land up to the margins of every shore. The island has seemed the perfect form in English cultural imagining, as the city was to the Greeks. Defensive, secure, compacted, even paradisal – a safe place; a safe place too from which to set out on predations and from which to launch the building of an empire. Even now, remote islands – the Falkland Islands or Fiji – are claimed as peculiarly part of empire history.

The island is equated with England in the discourse of assertion, though England by no means occupies the whole extent of the geographical island; Scotland and Wales are suppressed in this description and Ireland is corralled within that very different group, 'the British Isles'. In this century, of course, the disjuncture has become more extreme, with the division between Ulster and Eire, one assertively with 'the United Kingdom', the other an independent state. But Ireland has for far longer been the necessary other in the English description of England, 'John Bull's other island' which is determinedly *not* John Bull's.

Shakespeare's *Richard II* provided the initiating communal self-description, alluringly emblematic and topographical at once. Gaunt calls England:

> This royall Throne of Kings, this sceptred Isle.
> This earth of Majesty, this seate of Mars,
> This other Eden, demy paradise,
> This Fortresse built by Nature for her selfe,
> Against infection, and the hand of warre:
> This happy breed of men, this little world,
> This precious stone, set in the silver sea,
> Which serves it in the office of a wall,
> Or as a Moate defensive to a house,
> Against the envy of lesse happier Lands,
> This blessed plot, this earth, this Realme, this England.
>
> (*Richard II*, II, i, 42–52)

England is seen as supremely and reflexively *natural*: 'This Fortresse *built by nature for her selfe*'. The 'insularity' of the island is emphasized in this part of the speech: it is a 'little world', a moated country house as well as a fortress. It is both a miniature cultivated place, 'this blessed plot', and an extensive 'Realme'. It is 'this England' (as the *New Statesman* fondly titles its collection of symptomatic news-cuttings each week). The less-quoted second part of Gaunt's speech turns into an accusation against the present rate of this favoured island. First, he represents the noble fecundity of the land, 'this teeming wombe of Royall Kings', who are renowned for their deeds:

> This Land of such deere soules, this deere-deere Land,
> Deere for her reputation through the world,
> Is now Leas'd out (I dye pronouncing it)
> Like to a Tenement or pelting Farme.
> England bound in with the triumphant sea,
> Whose rocky shore beates backe the envious siedge
> Of watery Neptune, is now bound in with shame

The country house or fortress is become 'a Tenement or pelting Farme'. The binding in of the land by the sea – a natural battle of repulsion and attraction in which 'triumphant' curiously attaches both to land and sea and suggests a sustained and wholesome matching – is instead constrictingly 'bound in with shame'. It is easy to see why the latter part of this speech is less often ritually recalled than the first. Value ('this deere-deere Land') and price 'Leas'd out') are here set disquietingly close. Shame – the shame of bad government – mars the perfect order of the island.

The imagery which Shakespeare employs is defensive, not expansionist, though there is a suggestion of the depredations of the crusaders at the centre of the speech which speaks of the 'Royall Kings':

> Fear'd by their breed, and famous for their birth,
> Renowned for their deeds, as farre from home,
> For Christian service, and true Chivalrie,
> As is the sepulcher in stubborne Jury
> Of the Worlds ransome blessed Maries Sonne.

The punctuation in the first folio affirms that they are *renowned* 'far from home' rather than that their deeds took place far from home: but the ear receives both meanings. The passage has certainly been put to expansionist uses, though its insistence in context is on correcting the bad governance of the island so that its society may fulfil its demi-paradisal geography. In its later extracting from the play the passage has been repeatedly employed for self-congratulation rather than self-correction.

But the Shakespeare passage draws attention also to a fundamental tension in the idea of the island, one which has been to some extent concealed by the later phases of its etymology. The concept 'island' implies a particular and intense relationship of land and water. The *Oxford English Dictionary* makes it clear that the word itself includes the two elements: 'island' is a kind of pun. 'Isle' in its earliest forms derived from a word for water and meant, 'watery' or 'watered'. In Old English 'land' was added to it to make a compound: 'is-land': water-surrounded land. The idea of water is thus intrinsic to the word, as essential as that of earth. The two elements, earth and water, are set in play. An intimate, tactile, and complete relationship is implied between them in this ordering of forces. The land is surrounded by water; the water fills with shores. The island, to be fruitful, can never be intact. It is traceried by water, overflown by birds carrying seeds.[19]

The equal foregrounding of land and sea is crucial not only in understanding the uses of the concept in imperialism, but in the more hidden identification between island and body, island and individual. The tight fit of island to individual to island permits a gratification which may well rely not only on cultural but on pre-cultural sources. The unborn child first experiences itself as surrounded by wetness, held close within the womb. It is not an island in the strict sense since it is attached to a lifeline, an umbilical cord. It becomes an island, an isolation, in the severance of birth. Such conceptual power-sources are available for our speculation, even though they may not be directly represented. When Donne, in one of the most famous sentences in English, asserts that 'No man is an island', the words take their charge from their quality of paradox. They presuppose that the individual *is* ordinarily understood to be like an island.

In *The Tempest* Caliban's claim to the island condenses oedipal and land-descent discourses: '*by* Sycorax my mother'. The island is his progeny and his inheritance. It is also himself:

> This Island's mine by Sycorax my mother,
> Which thou taks't from me.

(The Tempest, I, ii, 391)

His claim to possession is matrilineal. He is ab-original.[20]

The island has features of the female body; the map of the British Isles has sometimes been represented as taking the form of an old crone. But England is only intermittently a woman in the symbolic discourse of the nation. Britannia is a considerable displacement of the island idea, though she carries Poseidon's trident for pronging the fish, and the foe. The sea, which encircles the land, can also bring enemies to its shores and occasionally, as when the Dutch sailed up the Thames in 1667, the island has been humiliated by foreign penetration. At such times the sexual imagery of invasion makes England for a while the 'mother-land' in the language of politics.

H.G. Wells astutely commented in 1927 on the contrast between the steamship and the aeroplane era: 'the steam-ship-created British Empire . . . is, aerially speaking, decapitated. You cannot fly from the British Isles to the vast dominions round and about the Indian Ocean without infringing foreign territory' (*The Way the World is Going*, p. 131). In the Victorian period, he suggested, the sea-tracks of the long-distance steam-ships could foster the illusion that the British Empire dominated the entire world, because it was possible to set out from the central island and stay always within either British or international waters. This is an ingenious rationalization of the expanionist phase of the island story. Since the sea is as important as the land for the island concept, the sea offers a vast extension of the island, allowing the psychic size of the body politic to expand, without bumping into others' territory: The aeroplane, on the other hand, though offering access to almost limitless space, must overfly the territories of other nations. It cannot imitate the extension of the island, magically represented by the silver pathways in the wake of ships, threads linking imperial England to its possessions overseas.

To the Lighthouse is Woolf's island story. The family group and the house are themselves contracted intensifications of the island concept: and, in a further intensification, the final separation of the individual each from each is figured in the work: 'We perish each alone', Mr Ramsay obsessionally recalls. The island is displaced, a Hebridean place oddly like St Ives in Cornwall where Virginia Stephen spent her childhood summers and where the harbour island was much painted by the St Ives group at about the same time that Woolf was writing her novel. Throughout the book, sometimes louder, sometimes muted, the sound of the waves is referred to. The sea is as much the island as is the land. The fisherman's wife, in the story Mrs Ramsay reads to James, longs for possession and for dominance, for control: that last wish is shared with Mrs Ramsay, and perhaps the other wishes too. 'That loneliness which was . . . the truth about things' permeates the book (p. 186).[21] The lighthouse itself is the final island, the last signifying object, amidst the timeless breaking of the sea: 'it was a stark tower on a bare rock', thinks James as they finally come close to the lighthouse in the last pages of the book. At the end of the book the First World War is over; the family is fragmented: the mother is dead, a son, a daughter; the fishes in the bottom of the boat are dead. Cam and James, looking at their ageing father reading, renew their silent vow to 'fight tyranny to the death'. But Cam's musing continues:

> It was thus that he escaped, she thought. Yes, with his great forehead and his great nose, holding his little mottled book firmly in front of him, he escaped. You might try to lay hands on him, but then like a bird, he spread his wings, he floated off to settle out of your reach somewhere far away on some desolate stump. She gazed at the immense expanse of the sea. The island had grown so small that it scarcely looked like a leaf any longer. It looked like the top of a rock which some big wave would cover. Yet in its frailty were all those paths, those terraces, those bedrooms – all those innumerable things.

Distance and retrospect is achieved at the end of *To the Lighthouse*: 'It was like that then, the island, thought Cam, once more drawing her fingers through the waves. She had never seen it from out at sea before' (p. 174). The long backward survey to the politics of Edwardian family life, to England before the First World War, which began to unravel through the image of the abandoned house in 'Time passes' here reaches conclusion: 'It is finished'. Lily's words – and those of the Cross – means also what they say. Things have come to an end. The period of empire is drawing to its close. The book ends; the picture is done; the parents' England is gone. In laying the ghosts of her parents Woolf clustered them within an island, a solitary island which no one leaves at the end of the book save to accomplish the short, plain journey to that final signifer, the lighthouse, whose significations she refused to analyse. The island is here the place of intense life and the conclusion of that form of life, both private and the image of a community from whose values she was increasingly disengaged.

To the Lighthouse is an elegy for a kind of life no longer to be retrieved – and no longer wanted back. In *To the Lighthouse* Woolf frets away the notion of stability in the island concept. The everyday does not last forever. The island is waves as well as earth: everything is in flux, land as much as sea, individual as well as whole culture. The last book of *To the Lighthouse* looks back at the conditions of before 1914. Implicit is the understanding that this will be the last such revisiting for the personages within the book.

The aeroplane

The absolute answering of land and sea to each other, which contributed ideas of aptness and sufficiency to the Victorians' understanding of 'England', will soon be disturbed by a change of axes: under water, in the air. Woolf is writing into the period at which the island could be seen anew, scanned from above. Her later writing shares the new awareness of island-dwellers that their safe fortress is violable. They look up, instead of out to sea, for enemies. Stephen Kern argues of the aeroplane:

> Its cultural impact was ultimately defined by deeply rooted values associated with the up-down axis. Low suggests immorality, vulgarity, poverty, and deceit. High is the direction of growth and hope, the source of light, the heavenly abode of angels and gods. From Ovid to Shelley the soaring bird was a symbol of freedom. People were divided in their response to flying; some hailed it as another great technological liberation and some foresaw its destructive potential. [22]

All felt its symbolic and its political power.

The old woman singing beside Regent's Park in *Mrs Dalloway* mouths a primal series of syllables which have persisted from the prehistoric realm, a sound composed out of 'the passing generations . . .vanished, like leaves, to be trodden under, to be soaked and steeped and made mould of by that eternal spring –

> ee um fah um so
> foo swee too eem oo. [23]

(p. 90)

The aeroplane in the same novel, writing its message in the air, seems at first equally unreferential. Earth and air, sound and sight, resist signification though not interpretation. The people of the book set to, reading their messages into the community and into private need. The aeroplane is sybaritic, novel, and commercial. Its intended message is nugatory: an advertisement for Kreemo toffee, but it rouses in the watchers, many of whom do not appear elsewhere in the novel, thought, pleasures, and anxieties both glancing and profound.

In *Mrs Dalloway* the aeroplane is set alongside, and against, the car. (Both are observed by most of the book's named and unnamed characters.) The closed car suggests the private passage of royalty, and becomes the specular centre for the comedy of social class: the 'well-dressed men with their tail-coats and their white slips and their hair raked back' who stand even straighter as the car passes; 'shawled Moll Pratt with her flowers on the pavement'; Sarah Bletchley 'tipping her foot up and down as though she were by her own fender in Pimlico'; Emily Coates thinking of housemaids, and 'little Mr Bowley, who had rooms in the Albany and was sealed with wax over the deeper sources of life' (p. 23). All these briefly named characters respond to 'some flag flying in the British breast' and gaze devotedly on the inscrutable vehicle whose occupant is never revealed. The sharp description, as so often in Woolf's representations of the English classes and their rituals, inches its way towards hyperbole. At White's:

> The white busts and the little tables in the background covered with copies of the *Tatler* and bottles of soda water seemed to approve; seemed to indicate the flowing corn and the manor houses of England; and to return the frail hum of the motor wheels as the walls of a whispering gallery return a single voice expanded and made sonorous by the might of a whole cathedral.

(p. 22)

The continuity of club and cathedral, of London institutions (White's and St Paul's), and of London typecast characters, mocks the self-esteem which expands from individual to nation, and which clusters upon an invisible and yet over-signifying personage inside the car.

In the next paragraph Emily Coates looks up at the sky. Instead of the muffled superplus of attributed meaning represented by the car, the aeroplane is playful, open, though first received as ominous. Its 'letters in the sky' curl and twist, offer discrete clues to a riddle whose meaning will prove trivial – and the writing teases the reader with the ciphering of 'K.E.Y.':

> Every one looked up.
>
> Dropping dead down, the aeroplane soared straight up, curved in a loop, raced, sank, rose, and whatever it did, wherever it went, out

fluttered behind it a thick ruffled bar of white smoke which curled and
wreathed upon the sky in letters. But what letters? A C was it? an E,
then an L? Only for a moment did then lie still; then they moved and
melted and were rubbed out up in the sky, and the aeroplane shot
further away and again, in a fresh space of sky, began writing a K, and
E, a Y perhaps?

'Blaxo', said Mrs Coates in a strained, awe-stricken voice, gazing
straight up, and her baby, lying stiff and white in her arms, gazed
straight up.

'Kreemo', murmured Mrs Bletchley, like a sleep-walker. With his hat
held out perfectly still in his hand, Mr Bowley gazed straight up. All
down the Mall people were standing and looking up into the sky.

(pp. 23–4)

Everyone's attention is distracted from the car which '(. . .went in at the
gates and nobody looked at it)'. The repeated word 'up' disengages the
people from society. In their gazing the whole world becomes 'perfectly
silent, and a flight of gulls crossed the sky . . . and in this extraordinary
silence and peace. . .bells struck eleven times, the sound fading up there
among the gulls'. The contemplative erasure of meaning accompanies the
wait for meaning. Instead of the expansion of the car's hum to cathedral
size, the plane produces a modest insufficiency of meaning, and
amalgamates with sky and gulls, its sound fading instead of resonating.
It becomes an image of equalizing as opposed to hierarchy, of freedom
and play, racing and swooping 'swiftly, freely, like a skater'. It includes
death, 'dropping dead down', the baby 'lying stiff and white in her
arms', but it does not impose it. Then, as in *Orlando*, 'the aeroplane
rushed out of the clouds again'. Each person reads the plane's message
differently. To Septimus 'the smoke words' offer 'inexhaustible charity
and laughing goodness'. The communality is not in single meaning but
in the free access to meaning. The ecstatic joke is about insufficiency of
import: 'they were advertising toffee, a nurse-maid told Rezia'. The
message does not matter; the communal act of sky-gazing does. For each
person, their unacted part becomes alerted: Mrs Dempster 'always
longed to see foreign parts' but goes on the sea at Margate, 'not out o'
sight of land!'. The plane swoops and falls; Mrs Dempster pulls on the
thought of 'the fine young feller aboard of it'. The eye of the writer now
expands the aeroplane's height '*over the little island of grey churches*, St
Paul's and the rest' until, with an easy and macabre shift of perspective,
the plane reaches the 'fields spread out and dark brown woods' (we are
still soaring visually), then the eye of the writing homes downwards in
magnification to 'where adventurous thrushes, hopping boldly, glancing
quickly, snatched the snail and tapped him on a stone, once, twice,
thrice'. The aggression of the aeroplane is displaced on to the bird.

The liberated, egalitarian extreme of the aeroplane's height, and the distanced eye of the writing, dissolves bonds and flattens hierarchies. The passing plane raises half-fulfilled musings in the thoughts of another momentary figure. Mr Bentley, 'vigorously rolling his strip of turf at Greenwich', thinks of it as:

> a concentration; a symbol. . . of man's soul; of his determination, thought Mr Bentley, sweeping round the cedar tree, to get outside the body, beyond his house, by means of thought, Einstein, speculation, mathematics, the Mendelian theory – away the aeroplane shot.
>
> (p. 32)

The comic and disturbing discrepancies now are between the tight 'strip of turf' swept in Greenwich and the yearning towards the newly insubstantial 'real world' of post-Einsteinian theory. So the aeroplane becomes an image of 'free will' and ecstasy, silent, erotic and absurd. It is last seen 'curving up and up, straight up, like something mounting in ecstasy, in pure delight, out from behind poured white smoke looping, writing a T, and O, and F' (p. 33). Toffs and toffee are lexically indistinguishable, farts in the wake of lark, of sexual rapture. Virginia Woolf's disaffection from the heavily bonded forms of English society often expresses itself paradoxically thus as affection and play – and in this novel, as in *Orlando*, the aeroplane figures as the free spirit of the modern age returning the eye to the purity of a sky which has 'escaped registration'.[24]

The aeroplane in *Mrs Dalloway* is no war-machine. Its frivolity is part of postwar relief. It poignantly does *not* threaten those below. It is a light aircraft, perhaps a Moth. The D.H. Moth first flew in 1925 and, as Gibbs-Smith puts it, 'heralds the popularity of the light aeroplane movement'. At this period the aeroplane could serve as an image of extreme individualism and of heroism, as well as of internationalism. The hooded pilot becomes, in the 1930s, a trope in the work of W. H. Auden and it may be that Woolf's response to the work of Auden and his associates, sketched in 'The leaning tower', is more intense than has yet been charted. Certainly, the question of the individual artist's responsibility to produce revolutionary change in society becomes the matter of an impassioned argument between Woolf and Benedict Nicolson towards the end of her life. The argument is conducted in letters written with the drone of enemy aircraft overhead and in the period leading up to the evacuation of British troops from Dunkirk: that evacuation is a last island story, in which the little boats sailed by fishermen and amateurs from England impose a forlorn mythic victory upon a ghastly defeat. In the late 1920s and early 1930s, however, the aeroplane suggests escape and aspiration in her work.

Women were among the pioneers of early air travel and exploration. In the early 1930s came two individual exploits which drew immense acclaim: Amy Johnson flew solo from England to Australia in a Moth (4–5 May 1930),and in 1932 came the first solo Atlantic crossing by a woman (Earhart in a Vega).[25] The first woman to become a qualified pilot had been Baroness de Laroche as long ago as 1909. The figure of the aristocratic woman escaping ordinary confines is powerful symbolically for Woolf at the end of the 1920s, fuelled by her love affair with Vita Sackville-West, and permitting, as in dreams, an identification which brings impossible freedoms within the range of the everyday.

Woolf was fascinated, too, in a mood between the sardonic and the obsessional, with the failed dreams of escape, the Daedalean claims of women: we see it, at this period, in her story of Shakespeare's sister in *A Room of One's Own*. We see it also in the macabre comedy of her description of 'the flying princess', crossdressed in purple leather breeches, whose petrol gave out on a transatlantic flight in 1927 and who drowned with her companions:

The Flying Princess, I forget her name, has been drowned in her purple leather breeches. I suppose so at least. Their petrol gave out about midnight on Thursday, when the aeroplane must have come gently down upon the long slow Atlantic waves. I suppose they burnt a light which showed streaky on the water for a time. There they rested a moment or two. The pilots, I think, looked back at the broad cheeked desperate eyed vulgar princess in her purple breeches & I suppose made some desperate dry statement – how the game was up: sorry; fortune against them; & she just glared; and then a wave broke over the wing; & the machine tipped. And she said something theatrical I daresay; nobody was sincere; all acted a part; nobody shrieked; Luck against us – something of that kind, they said, and then So long, and first one man was washed off & went under, & then a great wave came & the Princess threw up her arms & went down; & the third man sat saved for a second looking at the rolling waves, so patient, so implacable & the moon gravely regarding; & then with a dry snorting sound he too was tumbled off & rolled over, & the aeroplane rocked & rolled – miles from anywhere, off Newfoundland, while I slept at Rodmell, & Leonard was dining with the Craniums in London.[26]

Woolf creates the incongruities of disaster: the clipped inhibited speeches 'nobody shrieked; Luck against us – something of that kind, they said, and then So long'; 'the rolling waves, so patient so implacable & the moon gravely regarding'. The pernickety vengefulness of this description and the glamour of the seascape make of flight a foiled escape, equally from England and from the sea of death. Air and water alike place small social life out of its element. The flying princess seems like a haunting other imagination for her own fear of flying too high, as well as being a savage pastiche of aristocratic claims to dominance.

Even Woolf's initial working title for *The Waves*, 'The moths', may have had an additional resonance lost to our ears. Clearly it refers predominantly to those flying creatures, so like butterflies to amateur eyes, but so particularly phototropic that at night they cluster helplessly towards any light source, even it if burns them to death; but in 1925 the Moth aeroplane first flew. *Orlando*, published in 1928, recognizes the aeroplane as an emblem of modern life, along with the telephone and radio, the lifts. Going up in the lift in Marshall and Snelgrove, Orlando muses: 'In the eighteenth century we knew how everything was done; but here I rise through the air; I listen to voices in America; I see men flying – but how it's done, I can't even begin to wonder. So my belief in magic returns' (p. 270).

Accepting technology into everyday life renews the magical; explanation becomes unstable and unsought. On the book's last page Orlando's husband, Shelmardine, returns from his rash voyage 'round Cape Horn in the teeth of a gale'. In an invocation of ecstasy which is both euphoric and comic the moment of midnight approaches:

> As she spoke, the first stroke of midnight sounded. The cold breeze of the present brushed her face with its little breath of fear. She looked anxiously into the sky. It was dark with clouds now. The wind roared in her ears. But in the roar of the wind she heard the roar of an aeroplane coming nearer and nearer. 'Here! Shel, here!' she cried, baring her breast to the moon (which now showed bright) so that her pearls glowed like the eggs of some vast moon-spider. The aeroplane rushed out of the clouds and stood over her head. It hovered above her. Her pearls burnt like a phosphorescent flare in the darkness.
>
> (p. 295)

The plane 'hovers' like a bird, mingling erotic and hunting imagery: 'it stood over her head'. It hovers also like the spirit brooding creatively. Pearls and landing lights are here confused: 'Her pearls burnt like a phosphorescent flare' and there is no gap between sea and air, pilot and captain, bird and plane and man: 'It is the goose!' Orlando cried. 'The wild goose.' Time coalesces: the time of the fiction and the time of the hand concluding the writing of the fiction coincide. The book ends: 'And the twelfth stroke of midnight sounded; the twelfth stroke of midnight,

Thursday, the eleventh of October, Nineteen Hundred and Twenty Eight' (p. 295).

In this confluence the aeroplane is the central image, here conceived as individualistic, erotic and heroic. 'Ecstasy' is enacted as a brillant ricochet of ancient and immediate symbol, which lightly draws on pentecostal signs; 'in the roar of the wind she heard the roar of an aeroplane'. Sounds become tactile: 'breeze, brushed, breath'. The labials and fricatives, 'br' repeated, lightly mimic the rumble of the approaching plane. The first stroke of midnight 'brushed' her face. 'The aeroplane *rushed* out of the clouds'. The man descends through the clouds at the conclusion here, but then, Shelmardine is 'really a woman' and Orlando 'a man', in their initial recognition of each other.[27]

The heady pleasure of air travel probably remained the more intense in Woolf's imagination just because she never flew. In her late essay she described flying over London with a convincing ease and élan which mischieviously resolves itself into fantasy at the end of the piece. She had been in London under bombardment; she had looked up anxiously after Vanessa vanishing by light plane to Switzerland. The aeroplane gave a new intensity to the upward gaze and the downward thump. Woolf saw the plane always from the point of view of the island dweller, aware of the intimate abrasion of land and sea, that intimacy now disturbed by the new pastoral of the aeroplane – pastoral because so strongly intermingled with breezes and country sights, lying so innocently 'among trees and cows', but sinister, too, ab-rupting the familiar lie of the land, the ordinary clustering of objects:

Monday 26 January

Heaven be praised, I can truthfully say on this first day of being 49 that I have shaken off the obsession of Opening the Door, & have returned to Waves: & have this instant seen the entire book whole, & how I can finish it – say in under 3 weeks. That takes me to Feb. 16th; then I propose, after doing Gosse, or an article perhaps, to dash off the rough sketch of Open Door, to be finished by April 1st (Easter [Friday] is April 3rd). We shall then, I hope, have an Italian journey; return say May 1st & finish Waves, so that the MS can go to be printed in June, & appear in September. These are possible dates anyhow.

Yesterday at Rodmell we saw a magpie & heard the first spring birds: sharp egotistical, like [illegible]. A hot sun; walked over Caburn; home by Horley & saw 3 men dash from a blue car & race, without hats across a field. We saw a silver & blue aeroplane in the middle of a field, apparently unhurt, among trees & cows. This morning the paper says three men were killed – the aeroplane dashing to the earth: But we went on, reminding me of that epitaph in the Greek anthology: when I sank, the other ships sailed on.[28]

Woolf uses 'dash' three times in this diary entry. She will 'dash off the rough sketch of the Open Door'; she saw 'three men dash from a blue car' and, last, 'the aeroplane dashing to the earth'. Lateral, vertical, horizontal: all these are figured by the one word. The hand writes; the men run; the plane falls. Speed unites them – a speed allayed by the last allusion to the sea. Imaginatively sky and sea are akin; and pilots are still *aeronauts*: sky-sailors. Anne Olivier Bell's note reads:

> Mount Caburn is the bare down dominating the Ouse Valley on the far side of the river from Rodmell. The crashed aircraft was an Avro 40K from Gatwick aerodrome, where the three dead men were employed. 'I am the tomb of a shipwrecked man: but set sail, stranger: for when we were lost, the other ships voyaged on.' Theodoridas, no. 282 in book VII of *The Greek Anthology*, Loeb edition.

Woolf experiences a totalizing of experience: air, sea, land, death, and life. Suddenly she sees 'the entire book whole', not an island, yet a totality.

The island's identity depends on water. It is the sea which defines the land. Wave theory disturbed the land–sea antinomies: instead, over and under, inner and outer, stasis and flux, became generalized as motion. Thresholds and boundaries lose definition. Something of this can be read in *The Waves*, a book whose rhythmic life is the reader's only means of pursuit. Instead of the 'man clinging to a bare rock', which was Virginia Woolf's image for herself as writer in the summer she began to write it, this book engages with an imaginative scientific world in which substance is unreal, motion universal.

This does not render *The Waves* an apolitical novel, but its politics are in its refusal of the imposing categories of past narrative and past society, its dislimning of the boundaries of the self, the nation, the narrative. In *The Waves* Woolf pushed on to the periphery all that is habitually central to fiction: private love relationships, the business of government, family life, city finances, the empire. Each of these topics is, however, marked into the narrative so that we also *observe* how slight a regard she here has for them. Instead she concentrates, as she foresaw women writers must do, on 'the wider questions . . . of our destiny and the meaning of life', instead of on the personal and the political. She does this by reappraising the world in the light of wave theory and the popular physics of Eddington and Jeans. Eddington writes in 1927: 'In the scientific world the concept of substance is wholly lacking . . . For this reason the scientific world often shocks us by its appearance of unreality'. He opens his argument by asserting that 'the most arresting change is not the rearrangement of space and time by Einstein but the dissolution of all that we regard as most solid into tiny specks floating in a void'.[29] Waves in motion are all the universe consists in: sound waves, sea waves, air waves – but as Jeans also observes in *The Mysterious Universe*: 'the ethers and their undulations, the waves which form the universe, are in all probability fictitious . . . they exist in our minds'. Jeans is thereby led to privilege fiction or equalize it with the outer world: 'The motion of electrons and atoms does not resemble those of the parts of a locomotive so much as those of the dancers in a cotillion. And if the "true essence of substances" is for ever unknowable, it does not matter whether the cotillion is danced at a ball in real life, or on a cinematograph screen, or in a story of Boccaccio'. He concludes that 'the universe is best pictured . . . as consisting of pure thought'.[30]

Yet people drown and planes crash. The silver and blue aeroplane sits in a field intact, dead men invisibly inside it. The air crash is a new form of death, and *thanatos* had great allure for Woolf. Septimus Smith and Percival both die falling from a height. However, 'when I sank the other ships sailed on'.

Motion is eternal, but the new forms of experience brought by flight also sharply focus social and national change. If, at one extreme, there is no island, only waves, at the other extreme the geographical ideal of England becomes more poignant in the Europe of the 1930s. Daedalus and Icarus – artificer, aeronaut, and unwilling sky-diver – were perhaps, I have suggested, imaginatively provoked into the writing of Joyce and Auden by the coming of the aeroplane. The conclusion of *Portrait of the Artist* foresees no aircrash. But Auden's poem, 'The Old Masters', combines the Breughel image of Icarus falling through the air with attention to the unnoticeable disaster and its concurrence with the everyday. It is a poem which evades the allegorical and refuses to mark more than suffering and oblivion. Woolf's description of the crashed airmen in the field has the same blithe calm.

The island and the aeroplane

Woolf considered herself no patriot. On 29 August 1939 she wrote: 'Of course, I'm not in the least patriotic.' In January 1941, while she was revising *Between the Acts* she wrote in a letter to Ethel Smyth:

> How odd it is being a countrywoman after all these years of being a Cockney!. . .You never shared my passion for that great city. Yet it's what, in some odd corner of my dreaming mind, represents Chaucer, Shakespeare, and Dickens. It's my only patriotism: save once in Warwickshire one Spring [May 1934] when we were driving back from

Ireland, I saw a stallion being led, under the may and the beeches, along a grass ride; and I thought that is England.[31]

This passage occurs in a letter concerned with the repression of sexuality. We can gauge some of the counter-forces in Woolf's relations to the idea of England in the condensing of disparate elements within the remembered image: the invocation of Ireland as the necessary other island, the emphasis on maleness – the stallion – in the idea of England, and the sense of herself as exile. Only in London can she feel herself in kinship with the most 'English' writers, representing the phases of the literary canon (Chaucer, Shakespeare, Dickens).

The euphoric image of the aeroplane and the keen pleasure in its menace, which we have seen in her earlier responses, are set in a more difficult series of relations with the idea of island history in her novels of the later 1930s, *The Years* and *Between the Acts*. Virginia Woolf's insistence on her own 'unpatriotic' relation with England is nearly always formulated in relation to a concession. She resisted and deeply disliked the show of public mourning for the 'heroes' of the R101, lost in 1930 on an experimental flight from England to India – an attempt bound into imperialism and the wish to annex, beyond Wells's 'steamship empire'.[32] She disliked 'the heap of a ceremony on one's little coal of feeling': 'why should every one wear black dresses'. The sameness demanded by tight social forms always irritated her and roused her scepticism. Her use of plurals is a recurrent means of teasing island pomposity but it sometimes succumbs to a related social condescension: the opening of *The Years* employs the privilege of the narrative over-eye looking down on thousands of similar events. We begin with the sky: 'But in April such weather was to be expected. Thousands of shop assistants made that remark .. . . Interminable processions of shoppers . . . paraded the pavements . . .'.

> In the basements of the long avenues of the residential quarters servant girls in cap and apron prepared tea. Deviously ascending from the basement the silver teapot was placed on the table, and virgins and spinsters with hands that had staunched the sores of Bermondsey and Hoxton carefully measured out one, two, three, four spoonfuls of tea.[33]

Something odd and uneasy occurs in this writing with its mixture of Dickensian super-eye and the autocracy of the air, gazing *de haut en bas*. The aerial view affords a dangerous narrative position, too liberating to the writer and demeaning to those observed here. The levity of this socially bantering view of London is corrected in the 1917 episode with the brief account of an air-raid, unseen from the cellar where the characters finish their dinner and wait for a bomb to fall. It does not fall on them; the silence, the 'greenish-grey stone', the oscillating spider's web are all seized into the writing with intense reserve (pp. 313–4). The remembered episode re-emerges in the 'Present day' section when Eleanor looks up to where she saw her first aeroplane and muses on the degree of change it has brought, thinking how it first seemed a black spot, then a bird. Next she recollects the 1917 raid, and then her eye falls on 'the usual evening paper's blurred picture of a fat man gesticulating'. Eleanor rips the paper violently, shocking her sceptical niece who has been feeling superior. '"You see", Eleanor interrupted, "it means the end of everything we cared for". "Freedom?" said Peggy perfunctorily. "Yes", said Eleanor. "Freedom and justice"' (p.357). Within the scan of two pages of memory the aeroplane has changed from bird to collusive war instrument, part of the oppression operated by dictators.

The writerly pleasure in the plane's fantastic powers, so prominent in *Mrs Dalloway* and *Orlando*, is now sardonically viewed. In *Between the Acts*, and in the diaries and letters which accompany its composition, Woolf works urgently on the problem of the artist's presence in society and in England's history. Outside the book she is in passionate controversy with her nephew on the artist's responsibility to bring about revolutionary change in society. In a caustic letter of 13 August 1940 she defends Roger Fry against Nicolson's charge of inaction.[34] As she writes *Between the Acts*, from May to August 1940, some of the worst of the war is going on directly over her head. On 9 June she writes: 'The searchlights are very lovely over the marsh, and the aeroplanes go over – one, a German, was shot over Caburn, and my windows rattled when they dropped bombs at Forest Row. But it's like a Shakespeare song today – so merry, innocent and very English'.[35]

English literature and English weather form much of the material of *Between the Acts*. Despite her disclaimers and her sense of being the townie incomer, Woolf was clearly engaged and puzzled by English life in a quite new way at the beginning of the war. The Women's Institute asked her to produce a play for them; instead what she did was to write Miss La Trobe and *Between the Acts*.[36]

It proved harder to let go of the island story once it was under threat from invasion, and once it seemed that she and her friends might, through inertia, have contributed to its obliteration. In the excellent work that has been accomplished on the connections between *Three Guineas* and *Between the Acts*, critics such as Roger Poole and Sallie Sears have drawn attention to the connection Woolf makes between militarism bred in men through their education and the coming of the Second World War.[37] The novel itself offers a comedic threnody for an England which may be about to witness invasion, the final loss of 'freedom and justice', and the obliteration of its history. This sounds a solemn task, but that is not Woolf's way of either celebrating or disturbing. Within the work she

alludes to and fragments the canon of English literature; she records a
tight and antique village community in whose neighbourhood has
recently been built 'a car factory and an aerodrome'; she places at the
centre an ancient house, Pointz Hall; and she mimics the self-
congratulatory forms of village pageants, then so often held on Empire
Day:

> 'The Nineteenth Century'. Colonel Mayhew did not dispute the
> producer's right to skip two hundred years in less than fifteen
> minutes. But the choice of scenes baffled him. 'Why leave out the
> British Army? What's history without the army, eh?' he mused.
> Inclining her head, Mrs Mayhew protested after all one mustn't ask too
> much. Besides, very likely there would be a Grand Ensemble, round
> the Union Jack, to end with. Meanwhile, there was the view. They
> looked at the view.
>
> (p. 184)

Woolf surrounds the people of the book with the contours of historical
landscapes no longer perceptible to the naked eye. On the second page
we move from the conversation about the village cesspool to the new
forms of aerial observation. The bird, singing, dreaming of the
'succulence of the day, over worms, snails, grit' is, as in *Mrs Dalloway*,
linked in sequence with the aeroplane. Mr Oliver 'said that the site they
had chosen for the cesspool was, if he had heard aright, on the Roman
road. From an aeroplane, he said, you could still see, plainly marked, the
scars made by the Britons; by the Romans; by the Elizabethan manor
house; and by the plough, when they ploughed the hill to grow wheat in
the Napoleonic wars' (p. 8).

The aeroplane, in this opening of the book, allows history to surface in
the landscape and be seen anew. The acceptance of change, new use,
and continuity, of village inconvenience and incomings, allows the
inconsequent of middle-class life a homely poetry which is as close as
Woolf comes to affection. In this work the 'future is disturbing our
present (p. 100). Giles, returned from London, rages at the 'old fogies
who sat and looked at views' when the whole of Europe was 'bristling
with guns, poised with planes. At any moment guns would rake that
land into furrows; planes splinter Bolney Minster into smithereens and
blast the Folly. He, too, loved the view' (pp. 66–7). The swallows swoop
as they have done since the world was a swamp, are caught into
language as 'the temple-haunting martins' (instead of 'martlets' in the
half-stirred allusion to *Macbeth* in the last serene evening before violence,
just as this is the moment before war). Their recurrence seems to offer
assurance of continuity, perhaps as factitious in its way (there may soon
be no temples) as the *Times* leader of yesterday:

The swallows – or martins were they? – The temple-haunting martins
who come, have always come . . . Yes, perched on the wall, they
seemed to foretell what after all the *Times* was saying yesterday.
Homes will be built. Each flat with its refrigerator, in the crannied wall.
Each of us a free man; plates washed by machinery; not an aeroplane
to vex us; all liberated; made whole . . .

(p. 213)

But it is old Mrs Swithin who carries in her mind an awareness of the
prehistory of England, of a voluptuous primal world even before
England was an island; 'Once there was no sea', said Mrs Swithin. 'No
sea at all between us and the continent' (p. 38). Now Giles reads in the
morning paper of men shot and imprisoned 'just over there, across the
gulf, in the flat land which divided them from the continent' (p. 58). So
the lackadaisical conversations about how far it is to the sea from Pointz
Hall also become part of a general dislimning of securities.

The land shifts, the sea dries up, the impermeable island is a
temporary form within the view of geological time. English life and
language, on its shorter scale, is similarly impermanent, and here Woolf
uses the parody sequences of the pageant to point the shifting markers of
the island literary canon. Isa musing on the library 'ran her eyes along
the books. "The mirror of the soul" books were. *The Faerie Queene* and
Kinglake's *Crimea*; Keats and the *Kreutzer Sonata*. There they were,
reflecting. What? What remedy was there for her at her age – the age of
the century, thirty-nine – in books?' (p. 26). The pageant opens with a
small girl who pipes:

This is a pageant, all may see
Drawn from our island history,
    England am I

(p. 94)

The child sticks there, having forgotten her lines. The pageant scenes
that follow include traditional elements (Queen Elizabeth played by the
local shopkeeper) but these images are ruffled and undermined,
sometimes by chance events – the wind, the cows lowing – sometimes by
the plethora of language. Beneath this spume lie inalienable emotions:
Love. Hate. To them, Isa, in this book, adds, Peace.

In *Mrs Dalloway* the dallying light aircraft represented the reassuring
triviality of peace after the war, which is still melting and freezing the
consciousness of Septimus Smith. Here the aircraft, still mingled in
imagery with natural forms and with happiness, also presage the future:
a future that may not exist.

Nothing holds its full form for long: that is one reason why rhyme, which fleetingly hitches unlike together, is so prevalent in the language of the book. Isa, dreaming of her tenuous secret love, looks out of her bedroom window at her little boy George with the two nursemaids in the garden:

> The drone of the trees was in their ears; the chirp of birds; other incidents of garden life, inaudible, invisible to her in the bedroom, absorbed them. Isolated on a green island, hedged about with snowdrops, laid with a counterpane of puckered silk, the innocent island floated under her window.
>
> (p. 20)

The fragile imagined island of security is succeeded by the image of arousal, here figured as aeroplane:

> the words he said, handing her a teacup, handing her a tennis racquet, could so attach themselves to a certain spot in her; and thus lie between them like a wire, tingling, tangling, vibrating – she groped, in the depths of the looking-glass, for a word to fit the infinitely quick vibrating of the aeroplane propeller that she had seen once at dawn at Croydon. Faster, faster, faster, it whizzed, whirred, buzzed, till all the flails became one flail and up soared the plane away and away . . .
> 'Where we know not, where we go not, neither know nor care', she hummed. 'Flying, rushing through the ambient, incandescent, summer silent . . .'
> The rhyme was 'air'. She put down her brush. She took up the telephone.
> 'Three, four, eight, Pyecombe', she said.
> 'Mrs Oliver speaking . . . What fish have you this morning? Cod? Halibut? Sole? Plaice?'
> 'There to lose what binds us here', she murmured. 'Soles. Filleted. In time for lunch please', she said aloud.
>
> (pp. 20–1)

Isa's habit of rhyming is skeined through the work without any satirical commentary and at times the same habit moves out into the communality of gossiping voices. 'So abrupt. And corrupt. Such an outrage; such an insult; And not plain. Very up to date, all the same. What is her game? To disrupt? Jog and trot? Jerk and smirk? Put the finger to the nose? Squint and pry? Peak and spy?' (p. 213). The semantic cacophony of rhyme (auditory likeness without referential reason) suggests the reckless antiquity of the community. The forms of likeness are embedded in the sounds of the language, not in any reasoned relationships. The slippage between words and senses, and between separate units of speech, is constantly displayed in this work, where words collapse, reverse, become units of lexical play without sustained boundaries. In this passage, for example, the fish 'sole', effortlessly reverses in her next line of poetry into 'lose': '"There to lose what binds us here", she murmured. "Soles. Filleted"'. 'Filleted' takes up the sense of *binds* (a fillet is a ribbon which bonds the hair); that sense flies loose, so to speak, beside the utilitarian boneless 'filleted' fish. The sole/soul fugue is elaborated a few pages later. The fugitive lightness of this linguistic play risks being lost in any act of analysis such as that I have just offered. But such inversions and smudging of semantic bounds are essential to the work's sense of smothered crisis. It has in itself contradictory functions: it signals collapse and fragmentation. Yet it also celebrates the insouciant resilience of the English language and of literary history: Woolf's one form of patriotism.

Poems do not survive intact in memory but single lines are absorbed and adapted. Past literature permeates the work, but as 'orts, scraps, fragments'. The canon of English literature is no tight island but a series of dispersed traces constantly rewritten in need, so that, for example, William Dodge misremembers Keats's 'Ode to a Nightingale' and is not corrected. Instead of the expected 'Grand Ensemble: Army; Navy; Union Jack; and behind them perhaps . . . the church' (p. 209) the audience at the end of Miss La Trobe's pageant is offered mirrors and the present moment. That moment fills with rain – nature takes its part – but is followed by interrupting aeroplanes. Nothing remains intact, and, within the gossip another idea emerges: 'The very latest notion, so I'm told is, nothing's solid' (p. 232).

At the end of the pageant, composed as it also is of 'scraps, orts and fragments' of past writing, Miss La Trobe holds the mirrors up to the audience 'reflecting. What?' 'Book-shy and gun-shy', all that is left to them is landscape and world gossip. Recurrence knits the island past together: the swallows fly in to the barn each year from Africa. 'As they had done, she supposed, when the Barn was a swamp' (p. 123). 'Before there was a channel . . . they had come' (p. 130). The skeined-out interconnections of the work (Giles's youth in Africa, the flight of the swallows, '"Swallow, my sister, o sister swallow", he muttered, feeling for his cigar case', the scene of the reported rape even) call to mind *The Waste Land*, though the mood is closer to *Four Quartets*: 'History is now, and England'. In this work, however, 'the doom of sudden death' oppresses all the characters, because it threatens the whole history of England. No longer is the island a sufficient geometry, a sustaining autonomy. 'The future shadowed their present, like the sun coming through the many-veined transparent vine leaf; a criss-cross of lines making no pattern' (p. 136).

In the intervals of the pageant the talk is of coming war, and of old

roses, of refugees, and the falling franc, of the royal family and Queen
Mary's secret meetings with the Duke of Windsor, and of 'the Jews,
people like ourselves, beginning life again' (p. 143). Mr Streatfield, the
clergyman, tries to draw the pageant together into coherent message: we
act different parts but are the same, a spirit pervades beyond our own
lives, 'Surely, we unite?' These hopeful utterances lead to his
announcement of the collection for 'the illumination of our dear old
church'. Woolf's first readers would have felt the force of the irony here.
This is mid-June; by mid-September 1939 all illumination will be doused
and the blackout will be in place.[38] And, within the text, prompt on this
cue, Mr Streatfield hears what he at first takes to be 'distant music'. His
next words are severed. 'The word was cut in two. A zoom severed it.
Twelve aeroplanes in perfect formation like a flight of wild duck came
overhead. *That* was the music.'

Curiously, Woolf alludes irresistibly back here to the ending of *Orlando*:
the wild duck or wild goose, the perfect formation, momentarily
naturalize the aeroplanes. But in the ensuing pages, amidst the gossip,
we realized that the audience has recognized that ominous zoom-drone
music and what it portends:

Also why leave out the Army, as my husband was saying, if it's
history? *And if one spirit animates the whole, what about the aeroplanes?*

(pp. 230–1)

What we need is a centre. Something to bring us all together . . . The
Brookes have gone to Italy, in spite of everything. Rather rash? . . . *If
the worst should come – let's hope it won't – they'd hire an aeroplane they said.*

(p. 231)

*I agree – things look worse than ever on the continent. And what's the channel,
come to think of it, if they mean to invade us? The aeroplanes, I didn't like to
say it, made one think . . . No, I thought it much too scrappy.*

(p. 232)

Then when Mr Streatfield said: One spirit animates the whole – *the
aeroplanes interrupted*. That's the worst of playing out of doors.. . .
Unless of course she meant that very thing.

(p. 234)

Unless of course, she meant that very thing! The second ends: 'the
gramophone gurgled *Unity – Dispersity*. It gurgled *Un . . . dis* And
ceased.'

The stare upward at the aeroplanes in *Between the Acts* is written in
Woolf's diary in the same months as dread of invasion, and invasion by
parachutists. The skittishness of the plane in *Mrs Dalloway* has vanished

in Between the Acts. The twelve planes in perfect formation at the end of
Between the Acts are machines, though the pattern of their flight mimics
that of birds. The sombre untranslatability of the planes here is part of
the new meaning of the aeroplanes after the Spanish Civil War. Even as
she wrote, Leonard was on fire-watching duties and (15 May 1940)
'Behind that the strain: this morning we discussed suicide if Hitler lands.
Jews beaten up. What point in waiting? better shut the garage doors.' A
month later, on almost precisely the first anniversary of that 'mid-June
afternoon of 1939', in which *Between the Acts* sets its summoning of the
island past, she records Harry West's account of Dunkirk: a survivor's
tale in which the safe harbourage of the island still, amazingly, and
perhaps only momentarily, holds.

> It pours out – how he hadnt boots off for 3 days, the beach at Dunkirk
> – the bombers as low as trees – how no English aeroplanes fought . . .
> At Dunkirk many men shot themselves as the planes swooped. Harry
> swam off, a boat neared. Say Chum Can you row? Yes, he said, hauled
> in, rowed for 5 hours, saw England, landed – didnt know if it were day
> or night or what town – didnt ask – couldn't write to his mother – so
> was despatched to his regiment.
>
> (20 June 1940)[39]

The jarring within this account flings a further sardonic beam upon the
refusal to accord, or to set in hierarchical order, or contain, that marks
the writing of *Between the Acts*. Refusing to resolve is here not
irresolution, but assertion. In the new world of flight and war the old
axes are turned, the old geometries of the island giving way. Woolf
writes always as a civilian from *within* the island, even as she records its
disliming. The 'we' of *Between the Acts* is that of the English language, of
intertextual play, and mythologized English history, viewed with the
sceptical yearning eye of Miss La Trobe. The slow flux of land-shifts
described in the book repeatedly reminds the reader that islands are
formed, not originary: Mrs Swithin, 'thinking of rhododendron forests in
Piccadilly; when the entire continent, not then, she understood, divided
by a channel, was all one' (p. 13). Overhead, rupturing the reiteration of
island life, go the war planes. Woolf did not live through that war but
she recorded a tonic and satiric elegy for the island.

Notes

1. For a thorough account of the history of flight see Charles Harvard Gibbs-
   Smith, *Aviation: An Historical Survey from its Origins to the End of World War II*,
   2nd edn (London: HMSO, 1985).

2. SIGMUND FREUD, in 'On dreams', added a section on symbolism in the 2nd edn (1911): he there instances airships. *Standard Edition of the Complete Psychological Works of Sigmund Freud*, ed. James Strachey (London: Hogarth Press, 1953), vol. 5, p. 684; lecture 10 of *Introductory Lectures on Psycho-Analysis, Standard Edition*, vol. 15, p. 155.

3. *The Diary of Virginia Woolf*, ed. Anne Olivier Bell (London: Hogarth Press, 1982), vol. 4, p. 113. In July 1932 the *Graf Zeppelin* took passengers for a circuit tour of Great Britain.

4. GERTRUDE STEIN, *Picasso* (London: Batsford, 1938) pp. 11, 50.

5. VIRGINIA WOOLF, *Between the Acts* (London: Hogarth Press, 1941), p. 231. All further references are to this edition.

6. H. G. WELLS, *The War in the Air, Particularly how Mr Bert Smallways Fared* (London: Bell, 1908) pp. 243–4.

7. NAUM GABO, quoted in *St Ives 1939–64: Twenty-Five Years of Painting, Sculpture and Pottery*, ed. David Brown (London: Tate Gallery, 1985).

8. HARALD PENROSE, *British Aviation: the Ominous Skies 1934–39* (London: HMSO, 1980), p. 290. Penrose gives no source for this remark. The emphasis on Chamberlain's *flight* to Munich to treat with Hitler in 1938 may seem curious to us, to whom such diplomacy is everyday. His flight, however, marked the entry of the aeroplane as a diplomatic instrument.

9. H. G. WELLS, *Twelve Stories and A Dream* (London: Ernest Benn, 1927), p. 5 (first published 1903).

10. 'The Argonauts of the Air', in *The Plattner Story and Others* (Leipzig: Tauchnitz, 1900), pp. 46–7.

11. *The Way the World is Going: Guesses and Forecasts of the Years Ahead* (London: Ernest Benn, 1928), p. 124.

12. VIRGINIA WOOLF, *Collected Essays* (London: Hogarth Press, 1966), vol. 4, pp. 167–72.

13. GIBBS-SMITH, op. cit., p. 250.

14. JAMES JOYCE, *A Portrait of the Artist as a Young Man* (London: Jonathan Cape, 1916), p. 288.

15. W. H. AUDEN, *The English Auden: Poems, Essays and Dramatic Writings 1927–1939*, ed. Edward Mendelson (London: Faber, 1977), p. 237.

16. *Ibid.*, p. 257; JOHN FULLER, *A Reader's Guide to W. H. Auden* (London: Thames and Hudson, 1970), p. 127.

17. *Variorum Edition of the Poems of W. B. Yeats*, ed. Peter Allt and Russell K. Alspach (New York: Macmillan, 1957), pp. 328, 791.

18. *Collected Essays* (London: Hogarth Press, 1966), vol. 4, p. 173.

19. Since writing this essay I have been studying the ways in which the idea of the island has entered a number of scientific discourses, as well as political and literary ones, within the last 150 years. This section of the present essay overlaps briefly with a much longer paper, 'Discourses of the Island', in a collection of essays on literature and science, edited by Frederick Amrine (Amsterdam: D. Reidel Publishers, 1988).

20. For an excellent discussion of *The Tempest*, and of *Robinson Crusoe* in the context of Caribbean colonization see PETER HULME, *Colonial Encounters: Europe and the Native Caribbean 1492–1797* (London: Methuen, 1986). *Robinson Crusoe* was one of the works Virginia Woolf most admired. See the discussion of its relation to the inception of *To the Lighthouse* in JULIET DUSINBERRE, *Alice to the Lighthouse: Children's Books and Radical Experiments in Art* (London: Macmillan, 1987) pp. 276–7, 323.

21. Compare my 'Hume, Stephen, and Elegy in *To the Lighthouse*, *Essays in Criticism* (Oxford: Oxford University Press, 1984).

22. STEPHEN KERN, *The Culture of Time and Space, 1880–1918* (London: Weidenfeld and Nicolson, 1983), p. 242.

23. *Mrs Dalloway* (London: Hogarth Press, 1925), p. 90. All further references are to this edition. See, for an enlightening discussion of this passage, MAKIKO MINOW-PINKNEY, *Virginia Woolf and the Problem of the Subject: Feminine Writing in the Major Novels* (Brighton: Harvester, 1987).

24. *Between the Acts*, p. 30: 'Beyond that was blue, pure blue, black blue; blue that had never filtered down; that had escaped registration'.

25. GIBBS-SMITH, op. cit., p. 251.

26. *Diary*, vol. 3, pp. 154–5. Compare my 'The Body of the People in Virginia Woolf' in Sue Roe (ed.), *Women Reading Women's Writing* (Brighton: Harvester, 1987) pp. 102–3.

27. *Orlando* (London: Hogarth Press, 1928) p. 295. All further references are to this edition. MAUD BODKIN, *Archetypal Patterns in Poetry: Psychological Studies of Imagination* (Oxford: Oxford University Press, 1934) p. 307, read Shel as a fantasy of masculinity in Orlando's mind: 'It is over his head – the aeroplane having now supplanted the ship – that there springs up the winged wild thing by which the woman finds herself haunted and lured'.

28. *Diary*, vol. 4, p. 7.

29. ARTHUR EDDINGTON, *The Nature of the Physical World* (Cambridge: Cambridge University Press, 1928), p. 274.

30. JAMES JEANS, *The Mysterious Universe* (Cambridge: Cambridge University Press, 1930), pp. 79, 136.

31. *The Letters of Virginia Woolf*, ed. Nigel Nicolson (London: Hogarth Press, 1980), vol. 6, pp. 354, 460.

32. *Diary*, vol. 3, pp. 322–3.

33. *The Years* (London: Hogarth Press, 1937), pp. 1–2. All further references are to this edition.

34. See for example *Letters*, vol. 6, pp. 413. 419, 421.

35. *Ibid.*, p. 402. Caburn was part of the prospect from Rodmell and acted as something of an emotional barometer for her: see, for example, her complaining entry (*Diary*, vol. 3, p. 322) about Leonard Woolf's family which ends: 'It is the most miserable of days, cold and drizzling, the leaves falling; the apples fallen; the flowers sodden; mist hiding.' Compare also note 28 above.

36. *Letters*, vol. 6, p. 391

37. See, for example, ROGER POOLE, *The Unknown Virginia Woolf* (Cambridge: Cambridge University Press, 1978), pp. 216–31; SALLIE SEARS, 'Theater of War: Virginia Woolf's *Between the Acts*' in Jane Marcus (ed.), *Virginia Woolf: a Feminist Slant* (Lincoln and London: University of Nebraska Press, 1983), pp. 212–35.

38. 'At this very moment, half-past three on a June day in 1939' (p. 92); 'sitting here on a June day in 1939' (p. 208). Penrose, op. cit., p. 276, points out that 'The thirtieth anniversary of Handley Page Ltd was on 12th June (1939) – at that time the country was spending almost £2 million a week on aeroplanes.' Living as she did so close to what was then Gatwick aerodrome Woolf could not fail to be aware of the significance of the greatly increased air traffic in the later 1930s and its war menace. When she wrote the novel she was under the flight path of invasion – not now by sea, to be repelled from the island fortress, but by air, with the land below under threat from paratroops and bombs.

39. *Diary*, vol. 5, p. 297.

9 Woolf's Room, Our Project: The Building of Feminist Criticism*

Catharine R. Stimpson

Stimpson's piece is no more one to try to summarise than *A Room of One's Own*, from which it takes its cues for a kind of women's righting (I did write this by mistake, but I'll let it stand . . .) which won't stay fixed in one place, be it that of a stable authorial identity or the one assigned to women by the agency Woolf did not hesitate to call a patriarchy. In her punning, playful style, running around the many rooms which Woolf's narrator enters during the course of her lecture, Stimpson carries on in practice Woolf's own outrageous refusal to stick to one definite feminist position, as she rails against the confinements imposed from outside. She also demonstrates how subsequent developments in feminist theory have taken up and extended, consciously or not, many of the separate suggestions so liberally dispersed along the way of this text which continues to function as a suitably unstable basis for feminist critical constructions. And at the same time as Woolf herself is open to criticism, so she can also be drawn on as a future critic of feminist theories which seem to block off too much, to think that the answers might be single, simple or final.

This essay is an assay, a descriptive weighing of feminist criticism, and an 'I say', a practitioner's account of its controlled and uncontrolled substances. My text is Virginia Woolf's *A Room of One's Own*, which she began in 1928.[1] Woolf's room has become a project that draftily houses us. In her power, failures, and perplexities, she is a major architect and designer of feminist criticism.

In 1928 Virginia Woolf was forty-six years old. In two years, she had published two major novels: *To the Lighthouse* and *Orlando*. Her body had become the presence photographs now commemorate: the sculptured bones; the huge, liquid eyes that at once delve into and elude another's gaze. In the fall, she accepted an invitation to give two lectures at

* Reprinted from *The Future of Literary Theory*, ed. Ralph Cohen (New York: Routledge, 1989), pp. 129–43.

Cambridge University, for her a haunted site. She admired the women of Cambridge, but pitied them their poverty and destinies, to become 'schoolmistresses in shoals'. She noted, ironically, 'I blandly told them to drink wine & have a room of their own.'[2] Her husband, Leonard, accompanied her to one lecture; her lover, Vita Sackville-West, to the other. Woolf was an equal opportunity companion.

In 1929, Woolf issued her revised lectures. She delighted in producing. She once noted, 'I mean its the writing not the being read that excites me' (*Diary*, III, p. 200). Yet, she feared the consumption of her toil. Predictably, she was anxious about potential responses to *A Room*. The terms of this anxiety are symptoms of a larger cultural malaise as well. In the 1920s, Woolf could observe and absorb the effects of the dismantling of feminism as a political movement and the robing of modern heterosexuality as a social norm.[3] In 1928, she and Leonard were possible witnesses at the obscenity trial of Radclyffe Hall's polemic against homophobia, *The Well of Loneliness*. Indeed, *A Room* sardonically, if fleetingly, refers to Sir William Joynson Hicks, the Home Secretary who prosecuted and persecuted Hall and her publisher. Woolf worried that readers, especially men, would find *A Room* too strident, its author too identified with women. She asked if her tone was not too 'shrill' and 'feminine'. She warned herself that she would be 'attacked as a feminist and hinted at for a sapphist'.

Woolf, though, was logophilic, logging in the forests of language. Ultimately, faith in her text defeated fear. Her diary notes that she wrote it with 'ardour & conviction' (*Diary*, III, p. 262). The public's reception justified that faith. Though she continued to brood about male anger, she cheerfully recorded her sales figures. Although money, and the approval it signifies, calmed her qualms, her diary still records the internalization of dominant sexual ideologies. When she writes of *A Room's* success, she mentions the stimulus of her marriage. Before publication, her anxiety about being read as a writer was inseparable from anxiety about being read as a lesbian. After publication, relief at being read as a successful writer was inseparable from relief at being a successful woman, that is, married.

Woolf's swerving, dancing, fleeing sets of maneuvers between homosexuality and heterosexuality prefigure three centers of gravity in feminist criticism. The first is women's interests. To be a feminist critic is to react to and act for women and, at the very least, their place and space in culture. The second center is heterosexual interests. The sexual preference of many feminist critics is heterosexual. So are their professional settings, their classrooms and libraries. The third center is male dominance. Men still regulate many of those sexual practices, classrooms, and libraries. Moreover, every feminist critic must come to terms, if not to grief, with traditional philosophical and cultural schemes, which men have constructed and then, like so many Mr Goodwrenches, patched up.

Feminist criticism seeks a method that places the first two centers in creative oscillation, in theory, for theory, in practice, for practice, and cuts off the pull of the third. Serving actual women and visions of new sexualities, feminist criticism refuses to serve male domination. At its worthiest, feminist criticism analyzes sexism, not as an isolated system, but as a synecdoche for other hierarchies. Sexism is their partner and a part that stands for the whole mean mess. Because of this, the feminist look at sex and gender is a perspective that permits us to gaze at everything.[4] The feminist overhaul of sex and gender is an activity that can permit us to overhaul everything.

Obviously, restlessness and suspicion are the preludes for such perspectives, this activity. Feminist criticism is neither a cheerleader for things as they are, nor the Quaalude of discourse. *A Room* is an agitating series of gestures that forbids complacency, security, and premature intellectual closure. Its quick shifts in tone – from sardonic to blandishing, pragmatic to fanciful, tough-minded to tantalizing, declamatory to nuanced – tug at any stable relationship between narrator and reader. Skillfully, Woolf uses the darts of style to project a theory that must stimulate, appeal to, and justify discontent. However, in part because her stylistic choices have a clear theoretical and rhetorical purpose, *A Room* remains within the margins of literature. It swims towards, but never dives into, what Rosalind Krauss has called 'paraliterary space': 'the space of debate, quotation, partisanship, betrayal, reconciliation; but . . . not the space of unity, coherence, or resolution that we think of as constituting the work of literature . . .'.

Woolf retains the Play with her drama; the Author with her voices; the Argument with her criticism.[5]

The Author slyly begins by paraphrasing another's voice, that of a member of the audience that her Primary Narrator is about to address as a Lecturer. This Mr Bennett-like voice from the crowd is demanding some reliable association between any lecturer's subject and the announced title. 'But, you may say, we asked you to speak about women and fiction – what has that got to do with a room of one's own?' (p. 3). The Author further appears to diminish the Lecturer's authority through presenting her as a wee thing. She has not sought out the podium. On the contrary, she has politely responded to a request to talk about women and fiction.

However, Woolf, the Author, is parodying the cultural passivity that Western culture has assigned to women and that *A Room* will lacerate. Within a paragraph, the Primary Narrator/Lecturer will confidently seize on a radical methodology. Refusing to hand over a product, a 'nugget of pure truth', she will instead describe an intellectual process, an approach to truths. She will foreground means, not end. Significantly, her process is to tell a story, to invent. Lies, she declares proudly, will 'flow' from her lips. Fiction will wrap around, and wrap up, history and criticism. As Woolf blurs the cognitive domains of fiction and fact, she implies that statements about history and criticism are just that, statements. Because they are cultural erections, they can come down as blithely, or as dogmatically, as they went up. Because of power of political structures over culture, a statement is often what the state has meant.

Woolf's manipulations of the narrative voice also simultaneously destabilize and restabilize the text. The Primary Narrator's identity is mobile, fluid, expansive, indifferent to an audience's wish to pin a person to a name. During her tale, she encounters and then, becomes, a series of Marys: '. . . call me Mary Beton, Mary Seton, Mary Carmichael or by any name you please' (p. 5). These Marys do have their meaning. Mostly unmarried, they are also figures in an old English ballad.[6] Woolf is proclaiming and claiming, entertaining and entering, a genre that she will later feminize.

The story of *A Room* is 'about' the Narrator/Lecturer preparing her talk 'about' women and fiction, itself a quest for women's genres. Slyly, Woolf calls on the pastoral for the first episode in that adventure. Waiting to lunch at an Oxbridge men's college, like King's or Trinity, the Narrator is fishing on a river bank. This is not the landscape of 'The Waste Land', which Eliot, a friend of the Woolfs, had published six years earlier. For the feminist, nature is lusciously poetic; fruitful. Male culture is arid; barren. As the Narrator strolls towards her party, across a 'grass plot', among buildings, the guardians of male culture – a choleric Beadle, a librarian who looks like an angel of death – block her way. The librarian keeps her from seeing holograph manuscripts by Milton, a father of modern English culture, and by Thackeray, the grandfather of Woolf's own stepsister. Literarily, the Narrator is disinherited.

However, lunch is lovely, a luxury, an idyll. In contrast, dinner at an Oxbridge women's college, Girton, perhaps, is sparse: gravy soup, grey meat, dry biscuits, tattered cheese. Lunch has been the stage set for wonderful company and talk. Dinner is a grungy setting. Afterwards, a tutor, Mary Seton, will tell the Narrator about the arduous founding of the college in the nineteenth century. Woolf was no Marxist, but *A Room's* juxtaposition of a winey lunch and a watery dinner, and of the histories of the colleges in which they are cooked, acerbically dramatizes the importance of money to education and public culture. So doing, Woolf anticipates the revelation of the material contexts of writing of Marxist feminist critics.

The Narrator next speaks from London, the center of the British Empire, which has employed many of the best and the brightest of male Oxbridge graduates. She is doing research for 'Women and Fiction' in the

British Museum. Meditating again on the power of wealth, she thinks of her aunt, Mary Beton, whose legacy of £500 a year has given the Narrator the freedom to write, to think about money in the British Museum. Like Mary Wollstonecraft, Woolf connects economic dependence and psychological autonomy; economic dependence and subservience. As a result, the money matters more to the Narrator in her daily life than the political right to vote. Moreover, the legacy has passed from one woman to another. Small though it might seem, compared to the assets of aristocrats and financiers, the gift of one woman to another is empowering, not crippling.

Despite its resources, the Museum frustrates the Narrator. She is an active reader, but the pages on its shelves tell her nothing about 'Women' or nothing but tripe. Again like Wollstonecraft, in *A Vindication of the Rights of Woman*, the Narrator is appalled and disgusted by the libraries about women. She leads, and provides material for, the feminist inquiry into women as readers and the constitutive power of gender in reading.[7] In love with textuality, but not with her culture's preferred texts, the Narrator goes home. She will spend much of the rest of *A Room* in her own library. Women, Woolf implies, will rewrite those preferred texts only when they are within their own spaces. During the next two days, the Narrator reads some more – first a first novel, *Life's Adventure*, by a woman, Mary Carmichael. Reveling in starting point after starting point, the Narrator, in an apostrophe, directly addresses the third Mary and dwells on her first sentence, 'Chloe liked Olivia' (pp. 86–8). The Narrator is happily responding to new cultural relationships: between woman as writer and woman as subject, and, within the fiction that emerges, between women as friends. Creating the Narrator, Woolf is herself establishing such relationships: between Woolf as writer and her Narrator/Lecturer as subject, and, within *A Room*, her fiction, between the Narrator/Lecturer and Mary Carmichael.

Later feminist critics have praised the Woolf/Carmichael redemption of women's literary friendships. However, the verb, 'to like', is less erotically charged than the verb 'to love' or 'to desire'. As Cora Kaplan has shown, Wollstonecraft's feminist theory reflects her retreat from links between woman and passion, in part because the ideology she was rejecting had restricted women to the sphere of the passions. Woolf shrinks from links between women and lesbianism, in part because the ideology she was half-accepting kept women to the sphere of heterosexual passions. If Wollstonecraft denies sexuality because she feared reinforcing a powerful construction of women's nature, Woolf denies lesbian sexuality because she feared the power of a burgeoning construction of women's nature.

The next day, after *Life's Adventure*, the Narrator rummages through some male culture: a novel by Mr A.; some criticism by Mr B., both worthy boobies. Their stuff is honest, but egocentric. The narrative 'I', like the male actor in history, is a phallic tower that overshadows and blots out the reality that surrounds it. In contrast, Woolf's narrative 'I', which may serve as a model for the female actor in the future, is a sinuous observer who insinuates herself into reality. Indeed, the Narrator merges, near the end of *A Room*, with an earlier character, the Fernham scholar Mary Beton. Woolf writes, 'Here, then, Mary Beton ceases to speak' (p. 109).

   Woolf cannot end her tale with a sentence about women's silence. She returns to her first scene :the Narrator/Lecturer responding to an interlocuter. Psychologically, the Narrator has dramatized a process of alternation between fused and individual identities. Now a single ego, she begins to banter with her audience. She mocks the genre of the public speech. 'Here I would stop', she jokes, 'but the pressure of convention decrees that every speech must end with a peroration' (p. 114). However, the lecturer is too committed to her subject, and to her audience, to close with a silvery and negating laugh. In 1662, in *Female Orations*, Margaret Cavendish bravely took the same male-dominated form and pulled it apart so that it might speak from and for women. The Narrator/Lecturer tries a similar transformation. She gives her audience, and Woolf's reader, an inspiriting, uplifting peroration in which she imagines newly born women writers who will redeem all those other women whom history has stifled, suffocated, and gagged: '. . . she would come if we worked for her . . . so to work, even in poverty and obscurity, is worthwhile' (p. 118).

   Obviously, this play has a point, this narrative something about which to natter. Woolf has designed a general theory of women, gender, and culture. As Wollstonecraft had done in the late eighteenth century, as Simone de Beauvoir was to do in *The Second Sex* in the mid-twentieth century, as all feminist critics now do, Woolf analyzes women as cultural victims who have suffered from a double whammy. Her invention of Judith Shakespeare, William's sister, the tragic heroine of a short story within the Narrator/Lecturer's quest tale, is a brilliant parable of female deprivation. First, women have been kept from producing culture, from being representors. If they have wanted to do so, two ghosts have inhibited them: 'Milton's bogey', and Margaret Cavendish, almost his exact contemporary, the 'crazy' Duchess of Newcastle, an aristocratic bag lady, a second '. . . bogey to frighten clever girls with' (p. 65). Both figures masculinize cultural authority and stir up female anxiety if women grasp that baton. Milton personifies the male author; the Duchess the fate of the woman who dares to be like him. Woolf's metaphor is apt, progressive, and crafty. A bogey is scary, but the province of the bogey is the nursery, not the study. Like Mrs Ramsay, with her son James in *To The Lighthouse*, we can dissipate bogeys with reason, confidence and love.

The second whammy is that the representations of women that culture does permit are, at best, distortions. Woolf suggests that people must serve as the primary witnesses of their own experience. If they do not, ideology will bear false witness. Although *A Room* lacks the invigorating detail of later feminist accounts of 'male' representations, like Kate Millett's *Sexual Politics* in 1970, Woolf deftly satirizes the pictures Englishmen have made and hung of Englishwomen. Leaking from the emotional poles of fear and sentimentality, these images now stick on the rhetorical poles of insult and gush.

Woolf's satire is neither an emotional nor a rhetorical pole. Rather, her games, her parodies, are a median strip between an anger against men that she feels, and fears, and a sweet concern for men that she fears, and does not feel. *A Room* runs between two Victorian ways: the enraged laughter of Bertha Mason and the loving smile of the Angel in the Home. Contemporary feminist criticism distances itself from Woolf's repression of anger as firmly as Woolf distances herself from Charlotte Brontë's expression of it. In the early 1970s, two exemplary texts – Adrienne Rich's poem 'The Phenomenology of Anger' and Hélène Cixous's prose poem 'The Laugh of the Medusa' – created images of women who could not really speak until they gave voice and vent to rage.

Since *Sexual Politics* in 1970, the study of the patterns of the representation of women has increased in sophistication. In the early 1970s, such work tended either towards a sociological focus on 'stereotypes' and 'images', which the machinery of socialization would imprint on little girls and boys, or towards a Jungian focus on 'archetypes', which the collective unconscious would pass through to little girls and boys.[8] Then, in the 1970s, feminist critics began to employ, and to mix variously, three methodological tools: structuralism and semiotics, as their guide to large-scale sign systems; revisionary psychoanalysis, as their guide to the relationship of language to the unconscious; and Marxist theory, as their guide to ideology. If these tools can be awkward together, each helps to supplement the earlier catalogues of 'women in culture' with 'gynesis', the exploration of 'the feminine', a signifier that both men and women can trade in our symbolic contracts.[9]

 Woolf blames 'patriarchal society' for women's deprivation. When she focuses on patriarchy as a psychological structure, as an organization of desire and feeling, she presents men as greedy for validation and self-esteem. Without them, men cannot control others, at home or abroad; cannot stick up colonial empires. Men need women to feed their psychic appetites. If Freud argues that civilization demands the repression of the instincts, Woolf counter-argues that an uncivilized civilization demands the repression of women. Without the Existential vocabulary of de Beauvoir, Woolf also predicts the dialectics of gender domination of *The Second Sex*. In 1938, in *Three Guineas*, Woolf explicitly adapts a rude Oedipal theory to explain why so many men deny the identity of others in order to assert their own; why so many fathers amputate the limbs of wives, daughters, sons and servants in order to build up prosthetic phalli. The logic of *A Room* is more metaphorical. Women are the enhancing mirror into which men must gaze, but never step behind. Women are men's addiction. Without woman, '. . . man may die, like the . . . fiend deprived of his cocaine' (p. 36).

Woolf cannot dwell with deprivation alone. Feminizing an ethic of heroism, she asserts that women have defeated the patriarchy's efforts to rule their creativity as well as procreativity. Women have been neither blinded, deafened, nor silenced. Indeed, their place outside of male culture, their alienation, has given them a critical viewpoint; their constricted passages a perspective; their artlessness an art. Such a sense of hope, of trust in women's possibilities, has buoyed up feminist criticism. In 1974, Alice Walker's poignant essay, 'In Search of Our Mothers' Gardens', was to apply Woolf's insights to black women. Walker was generous. *A Room* has the racism that infiltrates feminist criticism. In a treacherous paragraph, immediately after the parable of Judith Shakespeare, Woolf's Narrator is flipping through images that express men's need to appropriate people, places, and things. Distinguishing women from men, she generalizes: 'It is one of the great advantages of being a woman that one can pass even a very fine negress without wishing to make an Englishwoman of her' (p. 52).

Carelessly, cruelly, the sentence ruptures 'woman' from 'Negress', granting 'woman' subjectivity and 'negress', no matter how fine, mere objecthood.

Far less distastefully, Woolf suggests three strategies for mapping women's creativity that have influenced women's studies. First, at the same time as the *Annales* school in France was establishing modern social history, she calls for a women's history based on the kind of sources social history was to use: dietary customs, house plans, 'parish registers and account book' (p. 47). Next, she outlines a more specifically literary history, at least for white Englishwomen. So doing, she is an early practitioner of what Elaine Showalter was, in the 1970s, to call 'gynocritics', the study of the styles, genres, values, relationships and intertextualities of women writers. Woolf's originary figure is neither a mystic nor a queen, but Aphra Behn, the first professional woman writer in English. Her exemplary genre is neither the letter nor the diary, but the novel, the form that provided women with an income. Once again, Woolf, the daughter of Wollstonecraft, fuses economic independence and the textualized female imagination.

In 1985, Sandra M. Gilbert and Susan Gubar codified a women's literary tradition in English in their massive Norton anthology.[10] Not surprisingly, their 'Preface' opens with a quotation from *A Room*. Like Woolf, Gilbert and Gubar have had their trashings and thrashings. The least plausible, most hostile voices make two unrelated claims: (1) That no one can separate out women writers as a group, that 'women writers' is a silly, empty category; and (2) That someone can separate out women writers as a group, but that Gilbert and Gubar have transmogrified every woman writer into a feminist. More sensitive voices also make two, more connected claims: (1) That Gilbert and Gubar have left out too many women of color; too many citizens of Commonwealth countries; and too many experimental writers;[11] (2) That the anthology, nevertheless, as a Norton anthology, re-establishes the theory of the canon. A female canon is worse than a male, because feminist criticism was to disarm the canon's brag that a few, select texts have a mandate from art and history.

Since the 1970s, the study of women writers has expanded wonderfully, especially work with women of color; colonized women; lesbians; and, to a degree, ethnic women. *A Room* does not foresee the diversity of women writers, feminist critical approaches, and literary histories. Virginia Woolf looks back at Aphra Behn. However, Doris Lessing, gazing back at Virginia Woolf and Olive Schreiner, doubts the validity of a women's tradition. Alice Walker helps to recover Zora Neale Hurston. Adrienne Rich looks across at Audre Lorde and back at Anne Bradstreet and Emily Dickinson. Feminist criticism has had to polish lenses other than those of gender to sight its subject: lenses of race, class, nationality, age, religion, sexuality, community culture. For example, does a writer hear an oral tradition or not?

No feminist critic can gloss over the conflicts these lines of sight engage. For women writers are more than women. They belong to many 'writing communities' at once.[12] The citizens of one may suspect and hate the citizens of another. Woolf realized the causes of some differences among women. Mary Carmichael wears '. . . the shoddy old fetters of class on her feet' (p. 92). However, unable to foresee the diversity of women writers, Woolf could not imagine a feminist critic's mind full with the task of tracing the networks and stresses this diversity sets up. Among the virtues of the Norton anthology is that it reminds such a critic that s/he must eventually choose some texts over others to present to a reader, in classrooms and elsewhere. Because our critique of the canon and our catalogue of women writers are still incomplete; because the differences among women are so volatile and difficult, feminist critics have not always said how they will make such choices and why. They cannot defer much longer.

Finally, Woolf's third strategy for the mapping of women's creativity is to ask for a reinterpretation, a decoding, of despised women and marginal utterances. She writes: 'When . . . one reads of a witch being ducked, of a woman possessed by devils, of a wise woman selling herbs, or even of a very remarkable man who had a mother, then I think we are on the track of a lost novelist, a suppressed poet, of some mute and inglorious Jane Austen' (pp. 50–1).

Each of these figures – the witch, the possessed (madwoman or hysteric), the healer, the mother – is central to feminist criticism as an emblem of a woman struggling to speak in one sublimated vocality or another. Moreover, the Narrator/Lecturer goes on, the unsigned text might be by a woman. She ventures to guess '. . . that Anon, who wrote so many poems without signing them, was often a woman' (p. 51).

In *A Room*, the dissolution of the self in anonymity is painful. The self ought to be present in the signature. Embedded in Woolf's distress is a liberal, humanistic commitment to the individual. Not only should an author be associated with her, or his, work; name with act. Neither law, nor custom, nor the opinions of others should fetter the self. Each of us is most authentic when free; most free when authentic. Such beliefs are consistent with an endorsement of the conscious human subject, its capacity to use language to reach meaning, and some historical narratives. Such beliefs also sustain the feminist critic who wishes to subvert the patriarchs who have obliterated, ridiculed, and condemned woman as subject and to redeem her.

Such critics work from many positions, within many genres. Read, for example, a poem that seems to allude to *A Room*. A woman painting a room of her own. She has red paint.[13] She fantasizes that her room is going to look dangerous, explosive, infernal, or, even worse, vulgar. Then, in a series of flashes, she associates her paint with that on Greek vases, and the vases, the amphora, with her uterus and ovaries. Sardonically, she mourns, 'Ours is a red land, sour/with blood it has not shed . . . .' Reduced to blood and the womb, women have then been read as such. Yet, the speaker is painting her room, controlling the 'red'. The poet is writing, controlling what will be 'read'. This permits the final line: 'Why not a red room?' The question asserts a will to language, the recreation of a trope. It calls out the troops against a standard reading of a trope, but not against standard interpretations of the operations of a trope.

Logically, however, Woolf's commitments are inconsistent with Woolf's style, with that blurring of fact and fiction. Her irony, metaphors, and shifts in narrative voice call into question the coherence of the human subject; the possibility of solid truths that language might boost, even if language does slip, slide, congeal, and freeze, like putty in an O-ring; and our trust in representational acts as anything more than a series of signifying practices. Because of its inventiveness, *A Room* foreshadows a strain of post-modernism;[14] because of its self-division, a conflict within feminist criticism about post-modernism.

Let me defame through condensation. Several alternatives collide together in this contested zone in feminist criticism. Critic A can defend 'traditional' notions of self, discourse and meaning. She can, more or less, rely on realism. She can comfort herself with the fact that most feminists in practical politics implicitly agree with her. Critic B can valorize the holes and fissures in the discourse of the powerful because the feminine rises from them. Sibyl's voice spirals up from her cave; from dark cracks in ancient boulders. Critic B brews poetry with radical feminist and lesbian politics. Critic C can absorb herself in post-structural theory and shrug off political questions. To her skeptics, she embodies what might happen to feminist theory if academics seize and carry it away from feminist activity. Critic D, who hears the voices of some of the smartest feminist critics, can reconcile elements from post-modernisms and feminisms.[15] Like feminism, post-modernism trusts differences; like feminism, post-modernism distrusts hierarchies. Even more rigorously than most sorts of feminism, post-modernism prosecutes ideological and representational codes. Unhappily, unlike all feminisms, post-modernism often seems to forget actual women – their breath and bone, grit and grandeur – in its infatuation with the 'crisis of the subject' and the 'feminine' as a pre-Oedipal discursive mode.[16]

Think of Woolf in her reading chair with papers from these critics. She might respect the seriousness of a Critic A; handle passion of a Critic B delicately. Woolf reflects that passion in *Three Guineas*. If a Critic C were to forget the common reader, Woolf's gift for mockery might flash about. If a Critic D were doing intellectually risky things, in agile and deceptively simple language, Woolf might begin to take notes. She might ask of them all what was best for women, and for women and writing, at the present moment.

For much of *A Room* is a practical guide to what women need in order to work productively, excitingly. Woolf's most quoted prescription is for the virtuous room, a place of solitude and privacy, less luxurious than a mansion, less crowded than a tenement, and those five hundred pounds a year, enough to pay the rent on a nice room and stock its shelves. Yet, she is aware of other needs as well. Psychologically, her writers must have patience and stamina. Emotionally, like all writers and artists, her women should get validation and support from others close to them. Socially, her women must be educated. Woolf suspects the myth of the untutored genius, of either sex, of any class. For genius at once soars beyond and absorbs tradition. Emily Dickinson is herself and Protestant hymns. Only systematic education, from a school, from a parent, can teach tradition. Despite Woolf's influence, some academic feminists, like many radical feminists outside of the academy, believe that if women were to speak, they would speak beautifully and truthfully. Feminizing a Romantic delight in the primitive and a Rousseauistic delight in the pre-social, such a notion applauds the apparently spontaneous text: the letter, the journal, the diary. So, too, did Woolf, but not because she thought they proved the existence of untutored genius.

Finally, Woolf's writers, in the present, need new themes. A modernist, she feared nodding repetitions. Tentatively, but not tremulously, *A Room* suggests what these new subjects might be: first, women's lives, the Chloes and Olivias, their friendship, their laboratories, and next, our common life, inseparable from the domesticity, the washlines and kitchens, where women have lived so long and hard. Paradoxically, the everyday will provide the raw material of innovation. Like Woolf, feminist theory and criticism have recuperated the value of the daily here and now. Breaking into and irrevocably altering established cultures, the writing of the everyday will, for Critic A, reveal women's historical experiences; for Critic B, an essential female experience; for Critic C, the promise of feminist post-modernism. As Teresa de Lauretis proposes:

> feminism defines itself as a political instance, not merely a sexual politics but a politics of experience, of everyday life, which later then in turn enters the public sphere of expression and creative practice, displacing aesthetic hierarchies and generic categories, and which thus establishes the semiotic ground for a different production of reference and meaning.[17]

Despite its ardent flair, the prophetic leaps and historical grounding from which these leaps are made, problems hold up and gird *A Room*. They distress and baffle feminist criticism still. Saying what 'feminism' means to 'feminist criticism' is simpler than saying what 'woman' does. Feminism offers an analysis of history and culture that foregrounds gender, its structures and inequities; a collective enterprise that foregrounds women's resistance to inequities; and Utopian visions of a different, and better, future that enables and ennobles that resistance. Six decades or so after *A Room*, three decades or so after *The Second Sex*, 'woman' and 'women' furnish feminist criticism with analytical material; members of that collective enterprise; and the jig and saw of puzzles.

The specific puzzle that *A Room* represents, but never resolves, is whether 'woman' refers to a genus, which nature constructs, or to a characterological genre, which societies and culture construct. Since binary oppositions place 'woman' with and against 'man', such a query links up with those concerning sexual difference itself. At once charmingly seductive and teasingly elusive, Woolf declines to say if she refuses to speculate about the causes of sexual difference because she might incriminate herself; because she fears the wilds of uncertainty; because no one really knows; or because no one will ever 'really know'. Taking refuge in conceit, the Narrator gambols: 'But these are difficult questions which lie in the twilight of the future. I must leave them, if only because they stimulate me to wander from my subject into trackless forests where I shall be lost and, very likely, devoured by wild beasts' (pp. 80–1).

Woolf's glissandos about sexual difference echo through a famous, packed passage about the *womanliness* of the woman writer in which the synergistic association of undecidability and nature occurs. Genderizing grammar, these long paragraphs isolate a 'man's sentence', which Samuel Johnson and Edward Gibbon wrote, and a female sentence, which Jane Austen and Emily Brontë dared to pioneer. If such a fundamental unit of grammar is inflected, as female or male, surely rhetoric and genre might be inflected, too. Surely, the building block shapes the building, be it arcade or dome. For some, Woolf's female sentence dissents from and criticizes male dominance. It '. . . has its basis not in biology, but rather in cultural fearlessness . . .'[18] True, but only partly true. For *A Room* further naturalizes, not only undecidability, but women's writing. The Narrator muses: 'The book has somehow to be adapted to the body, and at a venture one would say that women's books should be shorter, more concentrated, than those of men . . .' (p. 81). *This* fencing sentence both extends nineteenth-century theories of sexual difference and points toward late twentieth-century theories of 'writing from the body' and '*écriture féminine*'. *These* female textualities promise to be polymorphous, polydimensional, polysemous. They will help women, no longer the Pretty Polly Parrots of the patriarchy, to disrupt and escape from unilateral, unilinear linguistic armaments.

In part to explain her genderizing of grammar, Woolf throws away a seed of a sentence that feminist criticism has planted and nurtured: 'For we think back through our mothers if we are women' (p. 79). Woolf's subject and predicate fuse two domains that patriarchs have severed: the first is male, public, productive, and rational; the second female, domestic, reproductive, and arational. The immediate context seems to limit 'mothers' to literary predecessors. Woolf's mothers seem to be Aphra Behn and Jane Austen rather than Julia Duckworth Stephen. However, feminist critics have responded to the entirety of *A Room*. They have also swung along with the return to the figure of the mother that feminist theory made in the mid-1970s. Their inquiries have become a matrix of overlapping, overlipping questions.

Some questions, the more positivistic, ask about the appearance of cultural representations of the mother. Others concern the meaning of that figure, 'The Mother Tongue'. Is there a Mother Tongue that licks us into speech? If so, where would new grammarians find that speech? As

Julia Kristeva suggests, in avant-garde texts? As Mary Daly urges, in the journals of radical feminists? Still other questions, which psychoanalysis has forged, seek the influence of the maternal presence in the creation of culture. Does the son's need to rip himself away from the mother, a self-conscious repetition of the cutting of the umbilical cord, lead to the father's culture? How does the cultural daughter mediate between the father's language and the mother's language? Does she have, as Margaret Homans discovers in nineteenth-century women writers, two languages? That of the father and of the mother.[19]

As feminist criticism has matured, it has begun to extend its female genealogy from twin sets, sisters, and mother/daughter dyads to grandmothers, stepmothers, aunts, and cousins. Yet, Woolf also warns against too exclusive an interest in women. The same text that praises women's lives, that oracularly murmurs of female sentences, declares, sternly indifferent to the priggish demands of internal consistency: '. . . it is fatal for anyone who writes to think of their sex. It is fatal to be a man or woman pure and simple . . .' (p. 108). The woman writer's model is no longer the good mother, but the androgyne, 'woman-manly or man-womanly' (p. 108). The woman writer must be capaciously interested in everything but membership in an interest group. Forbidden to focus too compulsively on her own sex, the woman writer must avoid the polemic on behalf of her sex, the feminist text. Strictly judged, *A Room* might fail Woolf's own test for real literature, for 'poetry'. Implicitly, Woolf is setting up a series of equivalences, each a unit that balances both genders: the creative mind; the good marriage; and, perhaps, the ballot box after female suffrage. Explicitly, if Woolf has invented Judith Shakespeare as the heroine of loss, she casts William Shakespeare as the hero of gender integration. He rebukes Milton, the cultural patriarch.[20]

Woolf's theoretical contradictions when she writes about the 'woman question', spaciously defined, have their reflecting image in her life in 1928: in love with, perhaps in bed with, Vita Sackville-West; in love with, certainly at home with, Leonard Woolf. In cultural criticism, in psychosexual practice, Woolf stands simultaneously outside, beside, and inside the borders of heterosexuality and of her sex. These borders are as fluid, as subject to redrawing, as those of counties and countries. Woolf's stance, then, occupies many places; her position many points.

This multiplicity can be a metonymy for the cultural situation of women and, more precisely, of feminist critics. For one woman belongs to many classes at once. No class has the same relationship to political or cultural power. That, obviously, is why there are many classes. Despite male dominance, female tokens drop into the turnstiles of society and rush down the fast tracks on the other side. Queen Elizabeth I knew classical languages and gave state speeches. *A Room's* Narrator feels alienated when she walks down Whitehall. In *Three Guineas*, despairing of society, Woolf will imaginatively transform alienation into a political group, the perpetually marginal Society of Outsiders. She will separate herself from Wollstonecraft and the feminist theory that justifies women's place inside society and, most important for Wollstonecraft, inside reason and discourse. Nevertheless, *A Room's* Narrator is a daughter of the British Empire. The generous aunt who leaves her a legacy dies in a fall from a horse in Bombay.

Some of the strongest feminist criticism shows how far 'inside' one, or more, of the boundaries of power many women writers are; how much might divide them from a reader, female or male, who is 'outside' those boundaries. In the 1970s, despite Woolf's doubts, Charlotte Brontë was a magnetic figure for feminist critics; the fictive autobiography of Jane Eyre an arche-narrative. Yet, in the 1970s, Marxist feminist critics revealed the ideological assumptions about class in *Shirley*. In the 1980s, Third World critics have spoken of *Jane Eyre* as a 'cult text' that underwrites the 'axiomatics' of imperialism. Why must Bertha die if Plain Jane is to live?[21]

Showing that some women have had some power helps to erode the representation of all women as powerless, which a brand of feminist theory has mistakenly proposed. Fidelity to complexity respects women's lives. In addition, presenting the differences among women helps to erode the attraction of falsely universal descriptions of women and sexual difference. The commonality of the female becomes nothing more, nothing less, than an accident of birth, a chromosomal draw. This, in turn, can erode the general attraction of the falsely universal. These moves are the promise of a critical method I call 'herterogeneity'. Herterogeneity is the marking of differences among women, for themselves and as a way of recognizing and living generously with all but homicidal difference/s – among tongues and tests; tribes and territories; totems and some taboos.[22] Less globally, because feminist criticism is supple enough to mesh with other tools, from bibliographical notations to theoretical hermeneutics, it opens up criticism itself to the push and play of several languages.

The explorations of feminist criticism begin with the experiences of women and the functions of sexual difference. Herterogeneity goes on to sight the differences among women that body and community, place and history, have bred, and, often, inbred. Dwelling with these distinctions instructs women and women in the disabilities of monolithic thinking about 'woman', binary thinking about the duality of 'woman'/'man'. In such instruction is, if not salvation, at least a salutation of the Other, others, and othernesses. Is there no room for this?

Notes

1. VIRGINIA WOOLF, *A Room of One's Own* (New York: Harcourt, Brace and World, Harbinger Book, 1929, 1957). Hereafter cited in the text by page number. The best account of Woolf's influence on feminist critics is JANE MARCUS, 'Still Practice, A/Wrested Alphabet: Toward A Feminist Aesthetic', *Tulsa Studies in Women's Literature*, **3**, 1/2 (Spring/Fall 1984), 79–97. For other citations that reveal the range of Woolf's appeal, see LILLIAN S. ROBINSON, 'Who's Afraid of *A Room of One's Own?'*, *The Politics of Literature: Dissenting Essays on the Teaching of English*, ed. Louis Kampf and Paul Lauter (New York: Pantheon, 1972; Vintage, 1973), pp. 354–411; SHEILA DELANEY, 'A City, a Room: the Scene of Writing in Christine de Pisan and Virginia Woolf', *Writing Woman* (New York: Schocken Books, 1983), pp. 181–97; TERESA DE LAURETIS, *Alice Doesn't: Feminism, Semiotics, Cinema* (Bloomington, Indiana: Indiana University Press, 1984), p. 158; MARY JACOBUS, *Reading Woman: Essays in Feminist Criticism* (New York: Columbia University Press, 1986), pp. 27–40; ELIZABETH ABEL, *Virginia Woolf and the Fictions of Psychoanalysis* (Chicago: University of Chicago Press, 1989), Chapter 5.

   Making A Difference: Feminist Literary Criticism, ed. Gayle Greene and Coppélia Kahn (London and New York: Methuen, 1985), and *The New Feminist Criticism*, ed. Elaine Showalter (New York: Pantheon, 1985) survey current feminist criticism, especially in the West.

2. VIRGINIA WOOLF, *The Diary of Virginia Woolf*, Vol. III, 1925–30, ed. Anne Olivier Bell (New York: Harcourt Brace Jovanovich, 1980), p. 200. Hereafter cited in the text by page number.

3. See MARTHA VICINUS, *Independent Women: Work and Community for Single Women 1850–1920* (Chicago: University of Chicago Press, 1985), pp. 247–80.

4. This is the move MYRA JEHLEN advocates in the influential 'Archimedes and the Paradox of Feminist Criticism', first published in *Signs: Journal of Women in Culture and Society*, **6**, 4 (Summer 1981), 575–601. ELLEN MESSER-DAVIDOW advances a similar argument in 'The Philosophical Bases of Feminist Literary Criticism', a superb survey in *New Literary History*, XIX (1987–88), pp. 63–103. CORA KAPLAN, 'Pandora's box: Subjectivity, Class and Sexuality in Socialist Feminist Criticism', *Making A Difference*, pp. 146–76, reconciles two strains of seeing/acting within feminist criticism: socialist feminist theory, concerned with social and economic processes, and liberal humanist theory, concerned with the 'self', with psychological autonomy, desires, feelings. Kaplan has shaped my comments on Mary Wollstonecraft and Virginia Woolf.

5. I quote and paraphrase ROSALIND E. KRAUSS, 'Poststructuralism and the Paraliterary', *The Originality of the Avant-Garde and Other Modernist Myths* (Cambridge: MIT Press, 1985), pp. 292–3.

6. ISOBEL GRUNDY, '"Words Without Meaning – Wonderful Words": Virginia Woolf's Choice of Names', *Virginia Woolf: New Critical Essays*, ed. Patricia Clements and Isobel Grundy (London: Vision Press and Barnes and Noble Books, 1983), pp. 215–16, discusses the importance of these names. She points out that Woolf leaves unmentioned the fourth Mary of the ballad, the narrator herself, Mary Hamilton, about to be hung for the murder of the baby 'who resulted from her seduction by the king'. This absent, but present, Mary is, like Judith Shakespeare, a figure of female suffering. See, too, SUSAN GUBAR, 'The Birth of the Artist as Heroine', *The Representation of Women in Fiction:*

Selected Papers from The English Institute, ed. Carolyn G. Heilbrun and Margaret R. Higonnet (Baltimore: Johns Hopkins University Press, 1983), pp. 20–1.

7. Woolf anticipates Judith Fetterley's formulation of the 'resisting reader' and Jonathan Culler's question, 'How does one read as a woman?' For Culler, this is a theoretical vantage point; a biological fact that experience has privileged; and a reminder of the construction of identity, including that of the reader (*On Deconstruction: Theory and Criticism after Structuralism*, Ithaca, New York: Cornell University Press, 1982, paper 1983, pp.43–64). *Gender and Reading: Essays on Readers, Texts, and Contexts*, ed. Elizabeth A. Flynn and Patrocinio P. Schweickart (Baltimore: Johns Hopkins University Press, 1986), is the best inquiry into the theory and practice of women readers.

8. I have discussed this in my essay, 'Feminism and Feminist Criticism', *Massachusetts Review*, **24**, 2 (Summer 1983), 272–88.

9. I adapt the term from Alice A. Jardine, *Gynesis: Configurations of Woman and Modernity* (Cornell: Cornell University Press, 1985), one of the most brilliant examples of the newer study of representations of women.

10. SANDRA M. GILBERT and SUSAN GUBAR, *The Norton Anthology of Literature by Women* (New York: W. W. Norton and Co., 1985).

11. For a succinct analysis of some of the difficulties of the Norton anthology, see MARJORIE PERLOFF, 'Alternate Feminisms', *Sulfur 14*, v, 2 (1985), 132, 134.

12. HORTENSE J. SPILLERS, 'Afterword', *Conjuring: Black Women, Fiction, and Literary Tradition*, ed. Marjorie Pryse and Hortense J. Spillers (Bloomington: Indiana University Press, 1985), p. 258. Among the many recent studies are PAULA GUNN ALLEN, *The Sacred Hoop: Recovering the Feminine in American Indian Traditions* (Boston: Beacon Press, 1986); MARY V. DEARBORN, *Pocahontas's Daughters: Gender and Ethnicity in American Culture* (New York: Oxford University Press, 1986); DEXTER FISHER (ed.), *The Third Woman: Minority Women Writers in the United States* (Boston: Houghton Mifflin, 1980); DIANNE F. SADOFF, 'Black Matrilineage: The Case of Alice Walker and Zora Neale Hurston', *Signs*, **11**, 1 (Autumn 1985), 4–26; MARTA ESTER SÁNCHEZ, *Contemporary Chicana Poetry: A Critical Approach to an Emerging Literature* (Berkeley: University of California Press, 1985).

13. JUDITH RODRIGUEZ, 'Nu–Plastik Fanfare Red', *Penguin Book of Australian Verse* (Sydney: Penguin Books, 1972), pp. 428–9.

14. TORIL MOI, in *Sexual/Textual Politics* (London and New York: Methuen and Co., 1985), celebrates this Virginia Woolf.

15. An argument of this, which includes but is not limited to feminist criticism, is R. RADHAKRISHNAN, 'The Post-Modern Event and the End of Logocentrism', *boundary 2*, **12**, 1 (Fall 1983), 33–60. My essay, 'Nancy Reagan Wears A Hat', takes up these issues in more detail. Reprinted in my *Where the Meanings Are: Feminism and Cultural Spaces* (New York: Routledge, 1988).

16. NANCY K. MILLER, a major feminist critic, fully aware of structuralist and post-structuralist theory, speaks of the need for such resistance. She traces the woman writer staging the drama of subjectivity. However, Miller shows how that drama can take place outside of older, false universals without leading to new universals. In *Feminist Studies/Critical Studies*, ed. Teresa de Lauretis (Bloomington, Indiana: University of Indiana Press, 1986), p. 107.

17. TERESA DE LAURETIS (ed.), *Feminist Studies*, p. 10. ELIZABETH A. MEESE, *Crossing the Double-Cross: the Practice of Feminist Criticism* (Chapel Hill, North Carolina: University of North Carolina Press, 1986), also seeks connections among post-modern criticism, feminist criticism, and political acts.

18. RACHEL BLAU DUPLESSIS, *Writing Beyond the Ending* (Bloomington, Indiana: Indiana University Press, 1985), p. 33.

19. MARGARET HOMANS, *Bearing the Word* (Chicago: University of Chicago Press, 1986). SUSAN RUBIN SULEIMAN, 'Writing and Motherhood', *The (M)other Tongue: Essays in Feminist Psychoanalytic Interpretation*, ed. Shirley Nelson Garner, Claire Kahane, Madelon Sprengnether (Ithaca: Cornell University Press, 1985), pp. 352–77, lucidly surveys the connections among psychoanalysis, the figure of the mother, and writing.

20. I stress Woolf's androgyny more than a clever article that reads *A Room* as a covered '. . . manifesto for female difference'. FRANCES L. RESTUCCIA, '"Untying the Mother Tongue": Female Difference in Virgina Woolf's *A Room of One's Own'*, *Tulsa Studies in Women's Literature*, **4**, 2 (Fall 1985), 253–64).

21. GAYATRI CHAKRAVORTY SPIVAK, 'Three Women's Texts and a Critique of Imperialism', *Critical Inquiry*, **12**, 1 (Autumn 1985), special issue on '"Race", Writing, and Difference', ed. Henry Louis Gates, Jr, 243–61. See, too, CHIKWENYE OKONJO OGUNYEMI, 'Womanism: The Dynamics of the Contemporary Black Female Novel in English', *Signs*, **11**, 1 (Autumn 1985), 63–80. Gilbert and Gubar's Norton anthology reprints three novels: *The Awakening*, *The Bluest Eye* and *Jane Eyre*.

22. In his essay, 'Freedom of Interpretation: Bakhtin and the Challenge of Feminist Criticism', *Critical Inquiry*, **9**, 1 (September 1982), 45–76, Wayne C. Booth connects theories of textual differences with feminist inquiry. Bringing Bakhtin and feminist theory together is more and more common. See, for example, MYRIAM DÍAZ-DIOCARETZ, 'Sieving the Matriheritage of the Sociotext', paper at the conference, *The Difference Within: Feminism and Critical Theory*, University of Alabama, October, 1986, now a book of that title edited by Elizabeth Meese and Alice Parker (Amsterdam/Philadelphia: John Benjamins Publishing Company, 1989).

10 Penelope at Work: Interruptions in *A Room of One's Own**

PEGGY KAMUF

Like Mary Jacobus's piece, this one weaves in and out of materials that occupy very diverse places in what Kamuf calls 'our culture's poetic text'. It opens with the Penelope silenced by men, but pursuing her deviously interminable work of deferral by daily undoing, who becomes in effect an early woman writer, just as she is herself altered by the incursions of twentieth-century women, including Woolf. Kamuf looks in at Woolf's stagings of the interruption and exclusion practised upon women in *A Room*, then Michel Foucault makes a passing appearance to break in upon, and break down, the imaginary coherence of a Western philosophical subject. But there is something tenaciously and traditionally single-minded in Foucault's style of argument, which works in a more systematic mode than that of the women's indirect strategies. Kamuf suggests that there may be differences glimpsed, though never confirmed, through the double-edged movements, weaving to unweave, of a text like *A Room of One's Own*, never finished and never one.

. . . but always
I waste away at the inward heart, longing for Odysseus.
These men try to hasten the marriage. I weave my own wiles.

(*The Odyssey*, XIX, 135–7)

As so often throughout our culture's poetic text, one encounters in *The Odyssey* moments of abyssal self-representation when the poem tries to occupy a place in two different and mutually exclusive spheres, when it slips between representing something and being the something represented. One such moment occurs in Book I, where it happens to coincide with the first direct representation of Penelope. In fact, Penelope enters the scene of narration in order to interrupt it. In the passage to

* Reprinted from *Feminism and Foucault: Reflections on Resistance*, ed. Irene Diamond and Lee Quimby (Boston: Northeastern University Press, 1988). An extended version is published in *Signature Pieces: On the Institution of Authorship* (Ithaca: Cornell University Press, 1988).

which I refer, Telemachos and the suitors are gathered in front of the palace, where they are listening to 'the famous singer . . . [who] sang of the Achaians' bitter homecoming / from Troy'.[1] Penelope, who 'heeded the magical song from her upper chamber', is drawn down the stairs and, in tears, begs the singer to choose another song. At this point, Telemachos takes the floor, reproaches his mother for her intervention and says to her:

'Go therefore back in the house, and take up your own work,
the loom and the distaff, and see to it that your handmaidens
ply their work also; but the men must see to discussion,
all men, but I most of all. For mine is the power in this household.'
Penelope went back inside the house, in amazement.

(*The Odyssey*, I, 356–60)

Much later in the poem, at a crucial moment that prepares Odysseus's attack on the suitors, Telemachos again sends his mother out of the room, using almost the same terms but with one important difference. Instead of the poem or discussion, it is an instrument of force – Odysseus's famous bow – that Telemachos orders his mother to leave in men's hands.

'Go therefore back into the house, and take up your own work,
the loom and the distaff, and see to it that your handmaidens
ply their work also. The men shall have the bow in their keeping,
all men, but I most of all. For mine is the power in this household.'
Penelope went back inside the house, in amazement.

(*The Odyssey*, XXI, 350–4)

By means of this repetition, the poem establishes a connection between the art of storytelling and the practice of force. Both fall within a son's prerogative to exercise power in his household, the power to send women out of the room. If, however, a distribution of power and the sexes occurs here, it turns on the designation of woman's work as 'the loom and the distaff', the instruments of weaving and spinning. Both of these tasks supply the poet with endless metaphoric possibilities in this tale of men whose fate, for example, is 'spun with the thread at his birth' (VII, 198), where the storyteller can spin out a well-made tale and where cleverness weaves designs and deceptions. Thus, in a way that we have been taught to recognize,[2] the exclusion of the distaff from manly discussion is necessarily incomplete, since Penelope's work is set out as a kind of material support for the metaphorical field from which the poem draws its crafty designs and deceptive stories. But rhetorical repetition is not all that is working here to confound the distinction Telemachos would make. Power in the household is interrupted in quite another fashion by a woman's art.

Pressed by her household to choose a new husband, Penelope does not want to decide. Instead, she has given herself the tedious task of unweaving by night what she has woven during the day. It is not a terribly clever trick, nothing like saying 'No man' to the Cyclops Polyphemos, although perhaps that is what her unweaving means. In any case, it is a homelier remedy in a tight spot, which works even though her suitors, unlike Odysseus's Polyphemos, are perfectly able to see the tissue of her lies. Like a spider, she watches them fly into the web she has stretched across the entrance to the room in which she sits weaving. It is the same room she enters at night when others suppose her in bed. Here, then, is Penelope's great secret, what no man can see, for no man imagines her anywhere but in bed. It is this secret passage out of the bedchamber that allows Penelope to promise her bed and yet always defer the terms of the promise: no clever play on words, but rather a spatial and temporal shift between the two centers of her woman's life preserves Penelope's indecision. The suitors remain strangers to a woman's work that is never done, the tedium of the interior. As a result, their manly discussion is mystified by an obvious trick.

A Room of One's Own, the published text of lectures delivered at Newnham and Girton Colleges in 1928, begins with the question of its own title: 'But, you may say, we asked you to speak about women and fiction – what has that got to do with a room of own's own? I will try to explain.'[3] Likewise, the title 'Penelope at Work' – that is to say, the right to claim attention to whatever Penelope might have to say about Virginia Woolf – needs some explanation. Because authority here is a fiction, it can claim only the credit due the speculations of a common reader, in the sense that Woolf gives that notion in her two anthologies of critical essays, *The Common Reader*. I would add as well the other sense taken by the narrator of *A Room of One's Own*, when she sets aside a more systematic sounding of the depths, examining instead 'only what chance has floated to [her] feet' (p. 78).

I invoke Penelope in order to give a name to what is at work in a text like *A Room of One's Own*, although the phrase 'at work' already covers up in too purposeful a fashion the way in which such work entails as well its own undoing. I take Penelope as a shuttling figure in power's household, one whose movement between outside and inside, violence and poetry, the work of history and the unworking of fiction may allow us to frame one or two notions about the place of woman's art. This figure, moreover, may also serve to reformulate that other notion of woman's exclusion that always seems to arise whenever one takes up the question of power in stories and in histories. Finally, then, Penelope is the name I take in order to designate a conjunction of fiction in history in which a woman's text plots the place of its own undoing.

As already mentioned, *A Room of One's Own* opens with the question of its title. To provide an answer, the lecture's narrator introduces another fictional narrator ('"I", she writes, "is only a convenient term for somebody who has no real being"', p. 4), who proceeds to recount a series of events interspersed with a chain of literary analyses. Asked to explain, in other words, the narrator promises an answer once she is through spinning out her story. But this narrative sets out from a doubling back, or a crossing out, in which a meaning, a sense of direction, gets lost.

Having finally fished up an idea for her promised lectures on women and fiction, the narrator has set off at a rapid pace across Cambridge's campus, little heeding where her feet are taking her. Where she might have been going, however, no one can tell, because she is instantly called back to an order of distinctions that her thought had put aside in its unruly eagerness:

> Instantly a man's figure rose to intercept me. Nor did I at first understand that the gesticulations of a curious-looking object, in a cut-away coat and evening shirt, were aimed at me. His face expressed horror and indignation. Instinct rather than reason came to my help; he was a Beadle; I was a woman. This was the turf; there was the path. Only the Fellows and Scholars are allowed here; the gravel is the place for me. As I regained the path the arms of the Beadle sank, his face assumed its usual repose, and though turf is better walking than gravel, no very great harm was done . . . [However], what idea it had been that had sent me so audaciously trespassing I could not now remember.
>
> (p. 6)

This setback is itself soon forgotten and the narrator is led, through a series of rapid associations, to set her course for a certain college library where one might consult the manuscript of Milton's *Lycidas*. Once again, she is carried forward unconsciously, her bodily movement forgotten as one text leads to another until it is a question no longer about Milton but about a Thackeray novel that brings her to the door of the library. Once again, her unruly associations have transgressed a fundamental order, and the intertextual weaving is broken off when the narrator is recalled to the reality of her own unfitness in such a place:

> but here I was actually at the door which leads into the library itself. I must have opened it, for instantly there issued, like a guardian angel

barring the way with a flutter of black gown instead of white wings, a
deprecating, silvery, kindly gentleman, who regretted in a low voice as
he waved me back that ladies are only admitted to the library if
accompanied by a Fellow of the College or furnished with a letter of
introduction.

(pp. 7–8)

In its initial movement, then, the text describes a zigzag, a repeated
reversal of direction. From this angle, we may begin to see how *A Room
of One's Own* frames the question of women and fiction within the field of
an exclusion. What appears there is a contradiction like the one the
narrator discerns in the following passage: 'if woman had no existence
save in the fiction written by men, one would imagine her a person of
utmost importance . . . But this is woman in fiction. In fact, as Professor
Trevelyan points out [in his *History of England*], she was locked up,
beaten and flung about the room (pp. 44–45). The zigzag produced by a
reversal of sense is here more clearly coordinated with the contradiction
of fiction by history. And this zigzag intersects as well with the question
of the title: Is 'a room of one's own', in other words, an image, a
metaphor with which to call up the immaterial, the timeless, and the
imaginary defeat of power, or is it rather that which supports the
metaphor, the denotative foundation on which figurative space is
constructed? A place in history that exists therefore in social, political and
economic contexts? Or a place that transcends these limits much in the
way the narrator looks down upon the street activity from her study
window? How does *A Room of One's Own*, in other words, negotiate this
angle of contradiction?

The narrator defers these questions by posing another in their place, as
if she had found another use for Penelope's trick of leaving one room for
another, as if the promise she has made engages her to keep the passage
open between these two spaces, to let them interfere with each other.
Woolf's narrator, for example, cannot simply escape into the library from
a ruder reality; once there, she is drawn back into the rudest of scenes
where young women are 'locked up, beaten and flung about the room'.
Here, then, is another locked room within the first. The second enclosure
takes shape in the fully loaded bookshelves lining the walls. Having
locked women out of the library, history still rages at her from within.
The narrator runs into this locked door repeatedly in the British Museum;
even at home, in her own library, the violent encounters continue. Again
and again she is shown the door. Again and again anger flares as it did
when she was politely told she could not enter the college library. 'Never
will I wake those echoes, never will I ask for that hospitality again, I
vowed as I descended the steps in anger' (p. 8).

The narrator spins in the revolving door of the library. While anger

pushes her out, something else pulls her back in. That something else
has the force of forgetfulness – in its pull, one forgets one's place, one's
self. In this back and forth motion, the narrator is strung out between an
exclusion or negation of women and a forgetting of herself as woman.
Here, then, may be as well one space of woman's writing, which always
risks hardening into the negative outline of anger and thereby losing its
chance for forgetfulness. This is the sense of the encounter with
Professor von X, whom the narrator sketches as she reads his thesis, *The
Mental, Moral and Physical Inferiority of the Female Sex.*

> Whatever the reason, the professor was made to look very angry and
> ugly in my sketch, as he wrote his great book. . . . Drawing pictures
> was an idle way of finishing an unprofitable morning's work. Yet it is
> in our idleness, in our dreams, that the submerged truth sometimes
> comes to the top. A very elementary exercise in psychology, not to be
> dignified by the name of psychoanalysis, showed me on looking at my
> notebook, that the sketch of the angry professor had been made in
> anger. Anger had snatched my pencil while I dreamt.
>
> (pp. 31–3)

In this moment, the narrator has a view not only of the ugly face of the
historian but also of her own distorted features: 'My cheeks had burnt. I
had flushed with anger.' Yet these interceptions that snatch the pencil
from the hand and push thought off the path it was following always set
up the possibility of a new direction in which to proceed. When the
negations of history are made to turn on themselves, the door of the
library spins, setting the narrator in motion once again.

> All that I had retrieved from that morning's work had been the one fact
> of anger. The professors – I lumped them together thus – were angry.
> But why, I asked myself, having returned the books, why, I repeated,
> standing under the colonnade among the pigeons and the prehistoric
> canoes, why are they angry? And, asking myself this question, I
> strolled off. . . .
>
> (p. 33)

Through these deflections that turn a discourse back on itself, *A Room
of One's Own* defines a novel position in relation to the locked room of
history. That is, since women's history cannot be studied in the library, it
will have to be read into the scene of its own exclusion. It has to be
invented – both discovered and made up. As it spins around its promise
to decide on the place of woman's writing, this text *ravels* the crossed
threads of history and fiction. It ravels – which is to say it both *un*tangles,
makes something plain or clear, and *en*tangles, or confuses, something.

An alternative definition of the transitive verb 'to ravel' is (quoting from the decisive Oxford authority) 'to unravel'. Turning in the door of culture's most exclusive institution, Penelopean work blurs the line between historical prerogatives and fictional pretensions, always deferring the promised end of its labor, unraveling clear historical patterns at its fictional border.

In order to specify further this figure of the self-raveling text, one may turn to three different moments in *A Room of One's Own* where interruption marks the scene of writing. First, however, let us take a rather large detour whose only logic may be that of one text interrupting and unraveling another. The digression is proposed in order to step beyond a limited notion of interruption and thus a limited reading of Woolf's text. It passes through the work of Michel Foucault, most particularly his *Will to Knowledge (La Volonté de savoir)*. It might be useful to break into *A Room of One's Own* with Foucault's history of sexuality so as to point up the zigzagging fault lines in Woolf's speculations about woman's writing. Although the fault lines are quite plainly there, they can be too easily overlooked when this text is taken as a model authority for a critical practice that is content to go on making nasty caricatures of Professor von X, the nameless author and authority of masculine privilege and feminine subjection. The fault line beneath this sketch is the notion of sexual differentiation as a historical production that, if it has produced a privileged masculine subject, cannot also be understood as originating in the subject it only produces. To the extent, however, that one accepts seeing 'man' at the origin of his own privilege, then, one chooses paradoxically to believe the most manifest lie of 'phallosophy': that of man giving birth to himself as an origin that transcends any difference from himself. It is with just such a notion of production that Professor Foucault's history, for example, may interrupt whatever sketch we might make of privileged masculine subjectivity.

To resume very quickly: Foucault elaborates his history over against a certain Freudian–Marxian tradition that has consistently distinguished sexuality from the power mechanisms that repress it. According to this common notion, which Foucault labels 'the repressive hypothesis', power is structurally opposed to the anarchic energy of sexuality and functions to repress it, for example, by forcing conformity to the model of the monogamous heterosexual couple. The corollary to this hypothesis, therefore, is the value placed on sexual liberation as evidence of effective resistance to the bourgeois hegemony of power. Foucault, on the other hand, proposes that the repressive model of power is at best a limited and at worst a mystified one, in so far as it accounts only for negative relations and ignores the far more pervasive evidence of power's production of *positive* – that is, real – effects. He argues that, for at least two centuries in the West, power has maintained just such positive relations to sex and sexuality – sexuality, that is, is to a large extent produced by power – and these have progressively assumed a more important role as means for articulating power effects in the individual and society. All of which is why the various movements of sexual liberation need to be systematically reevaluated as instances also of the deployment of a will to knowledge, of power's articulating itself in the first-person confessional mode that also constitutes sexual identity. In an earlier work on disciplinary institutions (*Discipline and Punish*), Foucault gives an even clearer distinction of power in modern Western society as articulated in the various sciences of the subject, through the increasingly refined and differentiated techniques of identifying and classifying the 'I' of any discourse.

While one should hesitate to force Woolf's text into parallel with this analysis, one may at least accept seeing in it a background for a certain ambivalence. Woolf consistently sets the apparent political and social gains of a new women's consciousness over against the disturbing signs of an intensification of exclusive sexual identities, of sexually grounded subjectivity, and of subjectively grounded sexuality. What can emerge perhaps from the excursus into Foucault's history is another context within which to read *A Room of One's Own* as turning away from this historical preoccupation with the subject, closing the book on the 'I'. The gesture one can now read somewhat differently is that of the narrator when, near the end of her story and after leafing through the works of many women writers from Aphra Behn to her own contemporary Mary Carmichael, she takes one last book off the shelf. It is a novel by a certain Mr A. (whose initial, like the Professor's X, seems to stand for the whole alphabet of possible proper names). Quickly, however, she replaces it on the shelf because

> after reading a chapter or two a shadow seemed to lie across the page. It was a straight dark bar, a shadow shaped something like the letter 'I'. Once began dodging this way and that to catch a glimpse of the landscape behind it. Whether that was indeed a tree or a woman walking I was not quite sure. Back one was always hailed to the letter 'I'. One began to be tired of 'I'.
>
> (p. 103)

What our detour through the Foucauldian critique should allow us to see is that the power of this 'straight dark bar' to obliterate everything it approaches is not a power derived from the identity of a masculine subject to which the 'I' simply refers. Rather, the identification of subjects is already an effect of power's articulating itself on bodies, differentiating and ordering their intercourse.

Having noted this, however, what remains of Woolf's particular

187

critique of the patriarchal subject's historical privilege? Have we not passed over this aspect in order better to assimilate Woolf's text into the broader critique of the humanistic subject which is Foucault's project? Is it simply insignificant that the latter's analysis never interrogates the hierarchical opposition of the sexes as an important link in the deployment of power, while that distinction repeatedly forces itself on Woolf's thought, interrupting it, causing it to lose direction? Is there not, in other words, a sense in which *The Will to Knowledge* itself occupies a privileged space that knows no interruption?

Consider, for example, what one may call the narrator of *The Will to Knowledge*, the 'I' that assumes direction of the discourse's argument. Like the narrator of *A Room of One's Own*, this 'I' is 'only a convenient term for somebody who has no existence', it marks only a relative position in a discursive or textual network. Nonetheless, it is in a clearly different position. As we have seen, *A Room of One's Own* proceeds on the model of an interruption that forces the narration to deviate in some fashion, that intrudes with an effective, forceful objection to the momentary forgetting of a woman's identity. In *The Will to Knowledge*, on the other hand, it is the narration that defines other discursive procedures as 'deviations' and, compared to Woolf's narration, itself proceeds virtually free from distraction, since no one ever gets in its way with anything but spurious objections. To cite just one instance, it anticipates the particular obstacle to its progress that the Lacanian theory of desire might post, the theory, that is, of desire as constituted in and by, rather than against, the law. That theory, then, has already carried out a critique of ego psychology's repressive hypothesis, but its implications for a history of sexuality are opposed to those drawn by Foucault. One need not enter too far into the details of this debate in order to appreciate the discursive mode in which this objection is first formulated. Foucault writes:

> I can imagine that one would have the right to say to me: By referring constantly to the positive technologies of power, you are trying to pull off a bigger victory over both [Lacanian psychoanalysis and ego psychology]. You lump your adversaries together behind the figure of the weaker one, and by discussing only repression, you want to make us believe incorrectly that you have gotten rid of the problem of the law [which is constitutive of desire].[4]

While the 'I' will eventually respond to this objection, notice how in this moment (but there are many such moments)[5] the discourse imagines another position from which to address itself as 'you'. Is it any wonder the narrator is never at a loss for a reply? These interruptions of the narrator's pursuit of the analysis may be frequent, but they are never serious, since no figure appears there who, like the Cambridge beadle, has the position and the power to wave the narrator off the turf or to demand to see his permit to enter the library.

It is in this sense, at least, that a discourse like Foucault's can still retain a place in the privileged domain of patriarchal thought, a train of thought that has been trained, precisely, to think without interruption. And in a very important sense the privileged space in question is The Room of One's Own. These capital letters will refer us to the original room, the room properly named, the room of the Cartesian subject, where *Ego sum* is struck as an emblem bearing a proper name, taking up space the limits of which can be delineated and, perhaps most importantly, where the subject becomes one – both singular and whole. Michel Foucault is among those who have forced entry into this room so as to see what is going on there. In an appendix to the second French edition of *Histoire de la folie (Madness and Civilization)*, he writes that it is 'a peaceful retreat' to which Descartes's philosopher retires in order to transcribe the exercise of radical doubt. In that exercise, the subject of the meditation encounters an early 'point of resistance' in the form of the actuality of the moment and place of meditation: the fact that he is in a certain room, sitting by a fire, before a piece of paper. These conditions – a warm body next to a fire, writing instruments – are then taken by Foucault as synecdoches of the whole system of actuality, which the subject cannot be thought to lack and still be posited as the subject of a reasonable discourse. In the appended essay to which I refer, 'My Body, This Paper, This Fire', Foucault imagines that the meditating subject would have to reason as follows:

> If I begin to doubt the place where I am, the attention I am giving to this piece of paper, and of the fire's warmth which marks my present moment, how could I remain convinced of the reasonable nature of my enterprise? Will I not, by putting this actuality in doubt, make any reasonable meditation impossible and rob of all value my resolve to discover finally the truth?[6]

For Foucault, Descartes's place in the history of the Western episteme is so important because it situates the juncture of an exclusion – of unreason, of madness – with the seizure of material reality by the Subject of Reason. By means of that exclusion and that seizure, reality can be a quiet place in which to meditate on oneself.[7] However, when Foucault takes up the synecdochal figure 'My Body, This Paper, This Fire' as the title of his essay, he does so in order to reassert the abrogated claims of madness, to reassert, that is, the points of resistance to the elaboration of a reasonable subject. In a certain sense, these points provide leverage on the subject's discourse and give access to intrusion into it.[8] 'It was as if someone had let fall a shade. . . . Something seemed lacking, something seemed different. And to answer that question, I had to think myself out of the room' (p. 11).

Let us place this scene of a certain kind of intrusion into reason's discourse beside another that is imagined by Woolf's narrator. One will recognize a few reasons for doing so: the actuality of a scholar's meditation, a resistance, an intrusion – all are in play here. The narrator in this passage is spinning out her image of the great man of letters, seen not as he labors in the overheated library of Cartesian discourse, but rather in an idle moment. In fact, he has left the actuality of the library for another room.

> He [e.g., Johnson, Goethe, Carlyle, Voltaire, or any other great man] would open the door of drawing-room or nursery, I thought, and find her among her children perhaps, or with a piece of embroidery on her knee – at any rate, the centre of some different order and system of life, and the contrast between this world and his own . . . would at once refresh and invigorate; and there would follow, even in the simplest talk, such a natural difference of opinion that the dried ideas in him would be fertilised anew; and the sight of her creating in a different medium from his own would so quicken his creative power that insensibly his sterile mind would begin to plot again, and he would find the phrase or the scene which was lacking when he put on his hat to visit her.
>
> (p. 90)

A man of letters, a scholar, leaves his place by the fire in that quiet room and opens the door to a drawing room or nursery. There, the weary philosopher's work is supplemented by a 'different medium' and he is given to see 'the scene which was lacking' from the drama taking shape behind the other closed door. Notice that Woolf's narrator is both making up and making up for this scene. It has no place in the history and the biographies of great men which one may consult. It is thus invented, but to take the place of what is missing in the scholar's medium. In other words, the encounter with a supplemental difference takes place as fiction in history. Or, rather, it takes place in a mode that has as yet no proper name. Woolf writes:

> It would be ambitious beyond my daring, I thought, looking about the shelves for books that were not there, to suggest to the students of those famous colleges that they should re-write history . . . but why should they not add a supplement to history? calling it, of course, by

some inconspicuous name so that women might figure there without impropriety?

(p. 47)

When it acts to restore a missing scene in history's self-narrative, the narrative of the great man, Woolf's text catches history at a loss for words, interrupted in its train of thought. What is restored here, then, is not simply some unrecorded moment in the history of power but an interval, a hiatus, where that discourse has been momentarily broken off.

In order to figure such an interval or interruption, Woolf's text creates a passage out of the library and into another room. Let us briefly compare this passage to the one located by Foucault in the Cartesian scene of meditation. The subject of that meditation reappears in Foucault's essay just as he depicted himself, sealed in his heated study. Now, we could say that Foucault, unlike Woolf, simply finds no reason to imagine the philosopher wandering about from room to room at a loss. No doubt one would have to acknowledge that such moments occur, but it is reasonable for the historian of discourse to exclude them. Indeed, if one did not exclude them but allowed such idle fantasies to intrude, then it could hardly be called history that one is writing. Notice how, when it is considered in this manner, the reasonable omission reassembles the elements of the Cartesian subject's exclusion of its own madness. In this sense, at least, it constructs history by figuring only this comfortably situated position of power.

To return to the scene as it is imagined by Woolf's narrator: surely the interruption figured there is too quickly, too easily recuperated to the benefit of the suspended work. The great man is just taking a little break. Woolf's text, however, also figures two other sorts of interruption that are not so neatly resumed within the continuous work of history. Both are described as eruptions into the space of woman's work.

The first frames the nineteenth-century middle-class woman who, if she wrote, 'would have to write in the common sitting-room' (p. 69) as Jane Austen did and where, of course, 'she was always interrupted' (p. 70). The narrator quotes this passage from James Austen-Leigh's memoir of his aunt Jane: 'How she was able to effect all this is surprising, for she had no separate study to repair to, and most of the work must have been done in the general sitting-room, subject to all kinds of casual interruptions. She was careful that her occupation should not be suspected by servants or any persons beyond her own family party' (p. 70). To this the narrator adds: 'Jane Austen hid her manuscripts or covered them with a piece of blotting-paper.' Austen, in a recognizably Penelopeian fashion, undoes her work repeatedly so that it might continue. Each interruption blots out evidence of a fictional work and replaces it with the cover of domestic tasks.[9] The homely fiction of

191

domestic enclosure disguises the worldly fiction. That fiction is thus situated historically, materially. At the same time, however, a certain historical determination of woman's place is also seen to be conditioned by a fiction and based on a ruse that hides the contradiction of history.[10]

To understand some of the possible implications of this double hinging effect of interruption, what I am calling Penelopeian labor, one need only imagine that the weary scholar whom we earlier followed out of his study into a drawing room might have, without realizing it, walked in on someone like Miss Austen and found her 'with a piece of embroidery on her knee – at any rate, the centre of *some different order* and system of life.' The scholar's visit to this lady culminates in an inspiration which allows him to fill a gap in the discourse of reason, the discourse produced in a space of no difference, no interruption. By rewriting this familiar scene as we are suggesting, the phrase 'some different order' comes to imply a difference not only from the order that governs the scholar's work, but as well a difference from itself in so far as that piece of embroidery just may hide the text unraveling the domestic scene. The inspiring vision of difference, that representation that always implies an identity, is acted out as a mask for this other difference from itself, the difference within identity. The scholar is able to draw inspiration for his task because he believes he has glimpsed a scene other than the scene of writing, has caught sight of someone different, doing something else. Yet, because there may be a hidden text in the picture, it is perhaps someone much more like himself whom he has interrupted. The man of letters – historian, biographer, novelist, playwright, or literary critic – has failed to see himself as already represented in the room he has entered; it is precisely this blindness to his own reflection that induces a credulous inspiration for his work. Is he not, like one of Penelope's suitors, fooled by his eagerness to find her keeping the promise of her embroidery? What the text may thus display beneath its embroidered cover is a self-delusion, and in the very place, at the very moment that the scholar imagines for himself a way to fill a gap in the self's narrative. If history records the subject's delusion about its own identity, then fictions like Austen's and Woolf's restore to history the moments that precipitate such delusions, moments when difference can just be glimpsed before it disappears beneath a reassuring cover of familiar design.

All of this, of course, is quite fanciful speculation. Indeed, the little fiction about Jane Austen may be even more far-fetched than it appears, since at least one of Austen's recent biographers suspects that the whole description of the author hiding her manuscripts is apocryphal, at the very least an exaggeration. Despite her caution, however, this biographer cannot wholly avoid perpetuating the fiction, for she writes: 'I think this story . . . must be the happy later *embroidery* of Austen's nieces.'[11]

Nevertheless, the caution is well placed. Let us try to conclude on more solid ground by returning to the language of Woolf's text and yet another scene of interruption. The passage in question begins simply enough with the phrase 'One goes into the room', followed by a dash, a punctuated hesitation. This pause is just long enough to raise a question about the identity of the 'one' in the opening sentence. Then, having hesitated, the narrator goes on: 'but the resources of the English language would be much put to the stretch, and whole flights of words would need to wing their way illegitimately into existence before a woman could say what happens when she goes into a room' (p. 91). This sentence marks the limit, or threshold, of any lecture on women and fiction. Unlike the ease with which one can imagine the scholar walking into the drawing room or nursery, a woman enters the room in an unfamiliar, yet-to-be-written, even illegitimate mode. Clearly, for Woolf, such forced entry into the language will not simply substitute a feminine 'one' for a masculine. This becomes clear when, as the passage continues, Woolf shifts, without transition, from the question of the identity of the subject entering the domain of language to that of the many rooms one may enter.

> One goes into the room – but the resources of the English language would be much put to the stretch, and whole flights of words would need to wing their way illegitimately into existence before a woman could say what happens when she goes into a room. The rooms differ so completely, they are calm or thunderous; open on to the sea, or, on the contrary, give on to a prison yard; are hung with washing; or alive with opals and silks; are hard as horsehair or soft as feathers – one has only to go into any room in any street for the whole of that extremely complex force of femininity *to fly in one's face*. How should it be otherwise? For women have sat indoors all these millions of years, so that by this time the very walls are permeated by their creative force.
>
> (p. 91; emphasis added)

In effect, Woolf displaces the issue of the 'one' who enters the room by figuring in rapid succession a series of rooms to be entered, surveyed, plotted, described. But less obviously intervening here in the question of one's identity is the insistence of a form of self-interruption. By substituting the passive 'a woman's room is entered' for the active phrase 'one enters the room', this passage creates a disturbance on both sides of the threshold of subjectivity. And when the place of the feminine subject is abandoned in view of the multiple places of the 'complex force of femininity', then, retroactively and with a certain delay, it has become possible to begin to say what happens when a woman enters the room: in a word, femininity, already there, already at work, *flies in one's face*. We must try to hear this phrase – a figure of self-interruption – in both its

possible senses at once: to become overwhelmingly obvious and to
transgress flagrantly some law or rule. There is both a recognition and an
infringement of the place of a creative subject that is no longer or not yet
a 'one'. The feminine 'subject' is here constituted through illegitimate
intervention in the language: its 'one-ness' resides already in the other's
place(s), its unity derives retrospectively from an infraction that flies in
the face of the grammatical order of subject and predicate.

Far more radically than first imagined, *A Room of One's Own* can offer
refuge to no 'one', for the history, no less than the fiction, accumulated
there leaves the door open to intrusion. As we began by suggesting,
Penelope's clever labor is figured by and reiterates the cleverness of
Odysseus. The stories of their different exploits together assemble the
elements for a meaningful reunion. In that fictional moment that closes
the circle of the poem, when the ruse of power rejoins the ruse of no
power, it has become impossible and thus irrelevant to know who is
interrupting whom, whose task is suspended and whose continues, or
which room is being entered and which left behind. Interpreted as a
space of interruption, *A Room of One's Own* cannot give title to the room it
names in its title. No 'one' figures there who is not already many and no
ownership guarantees there an undivided property. Instead, the title
promises a place of intermittent work, a book that, like a woman's
thought, a woman's body, is frequently broken in upon. And broken off.
We can leave the last word to the narrator who advises the audience at
her lecture that 'the book has somehow to be adapted to the body, and at
a venture one would say that women's books should be shorter, more
concentrated than those of men, and framed so that they do not need
long hours of steady uninterrupted work. For interruptions there will
always be' (p. 81).

Notes

1. Translated by Richmond Lattimore (New York: Harper, 1957), pp. 325–7; other references are noted in parentheses.

2. By, for example, J. HILLIS MILLER in 'Ariachne's Broken Woof', *Georgia Review* **30** (Spring 1977).

3. VIRGINIA WOOLF, *A Room of One's Own* (New York: Harcourt, Brace and World, 1929), p. 3; future references are noted in parentheses in the text.

4. MICHEL FOUCAULT, *La Volonté de savoir* (Paris: Gallimard, 1976), p. 108; my translation.

5. Perhaps the most striking example of the technique is the final section of *L'Archéologie du savoir* (Paris: Gallimard, 1969), where the discourse, in effect, interviews itself and answers all the questions it can think of.

6. MICHEL FOUCAULT, *Histoire de la folie à l'âge classique* (Paris: Gallimard, 1972), pp. 595–96. For a critique of Foucault's reading of Descartes, see JACQUES

DERRIDA, 'Cogito et histoire de la folie', in *L'Ecriture et la différence* (Paris: Seuil, 1967).

7. See SUSAN BORDO, 'The Cartesian Masculinization of Thought' (*Signs*, **11**, 3) for another, significantly different account of 'masculinization' as an effect of separation.

8. However, as Foucault writes in *La Volonté de savoir*, points of resistance 'by definition . . . can only exist in the strategic field of power relations' (p. 126). Jean Baudrillard has pointed out that resistance has a rather unexplained status in Foucault's discourse; see *Oublier Foucault* (Paris: Galilée, 1977), pp. 50ff.

9. On how this 'cover story' may be functioning thematically in Austen's novels, see SANDRA GILBERT and SUSAN GUBAR, *The Madwoman in the Attic* (New Haven, Connecticut: Yale University Press, 1979), pp. 153ff.

10. Woolf's tampering with the distinction between fiction and history should also be read as an effect of their mutual implication in each other. For an excellent study of this question, see SUZANNE GEARHART, *The Open Boundary of History and Fiction: A Critical Approach to the French Enlightenment* (Princeton, New Jersey: Princeton University Press, 1984).

11. JANE AIKEN HODGE, *Only a Novel: The Double Life of Jane Austen* (New York: Coward, McCann and Geoghegan, 1972), p. 133; emphasis added.

Notes on Authors

ELIZABETH ABEL, who edited the anthology *Writing and Sexual Difference* (1982), initially an issue of the journal *Critical Inquiry*, is the author of *Virginia Woolf and the Fictions of Psychoanalysis* (1989). She is Associate Professor of English at the University of California, Berkeley.

ERICH AUERBACH, who died in 1957, taught Comparative Literature at Yale; he is most famous as the author of *Mimesis: The Representation of Reality in Western Literature* (1946), from which 'The Brown Stocking' is taken.

GILLIAN BEER is Professor of English at the University of Cambridge. Her books include *Darwin's Plots: Evolutionary Narrative in Darwin, George Eliot, and Nineteenth-Century Fiction* (1983) and *Arguing with the Past: Essays in Narrative from Woolf to Sidney* (1989). She has edited *Between the Acts* for Penguin and *The Waves* for the World's Classics series.

RACHEL BOWLBY, who teaches at the University of Sussex, is the author of *Just Looking* (1985) on femininity, shopping and novels at the turn of the century, and of *Virginia Woolf: Feminist Destinations* (1988). She has edited *Orlando* for the World's Classics series, and two volumes of Woolf's essays for Penguin: *A Woman's Essays* and *The Crowded Dance of Modern Life*.

FRANÇOISE DEFROMONT is the author of *Virginia Woolf: Vers la maison de lumière* (1985) and a forthcoming book on Katherine Mansfield, *L'Aloès en fleur*. She teaches English and American literature at the Université de Paris, VIII (Vincennes).

TONY INGLIS teaches English at the University of Sussex. He has edited a Penguin selection of D. H. Lawrence's essays from *Phoenix*, and also the Penguin edition of Scott's *The Heart of Midlothian*.

MARY JACOBUS, Professor of English at Cornell University, is the author of *Reading Woman: Essays in Feminist Criticism* (1986), as well as of books on Wordsworth; she also edited one of the earliest collections of feminist criticism, *Women Writing and Writing about Women* (1979).

PEGGY KAMUF is the author of *Fictions of Feminine Desire: Disclosures of Heloise* (1982) and *Signature Pieces: On the Institution of Authorship* (1988). She is the editor of *A Derrida Reader: Between the Blinds* (1991), and teaches French and Comparative Literature at the University of Southern California.

SUSAN M. SQUIER is the author of *Virginia Woolf and London* (1985), editor of *Women Writers and the City* (1984), and co-editor of a book on war and gender in literature, *Arms and the Woman* (1989). She is Associate Professor of English at SUNY, Stony Brook, and is editing Woolf's *The Years* for Basil Blackwell.

CATHARINE R. STIMPSON is Professor of English at Rutgers University, and editor of the feminist series 'Women in Culture and Society' published by the University of Chicago Press. She has written extensively on feminist theory; some of her essays are collected in *Where the Meanings Are* (1988).

Further Reading

So much has been published on Woolf in the past twenty years that it is impossible to construct a list which might hope to do more than provide a reasonable number of leads in the main directions that have been pursued. I have not attempted to divide the criticism into categories such as feminist, deconstructive, Marxist, socio-historical, psychoanalytical, narratological, and so on, because in almost every case the text would fall into at least two of these, so that confinement in one or the other would be more misleading than helpful. Brief comments have been added either where the title does not give enough indication of the topic, or when some other note seemed useful.

Background source materials

BELL, ANNE OLIVIER and ANDREW MCNEILLIE (eds). *The Diary of Virginia Woolf*. 5 vols. 1977–84; repr. Harmondsworth: Penguin, 1979–85.

BELL, QUENTIN. *Virginia Woolf: A Biography*. 1972; repr. London: The Hogarth Press, 1982. (To date the definitive biography, written by Woolf's nephew.)

DESALVO, LOUISE and MITCHELL LEASKA (eds). *The Letters of Vita Sackville-West to Virginia Woolf*. New York: William Morrow and Company, Inc., 1985.

GORDON, LYNDALL. *Virginia Woolf: A Writer's Life*. Oxford: Oxford University Press, 1986.

KIRKPATRICK, B. J. *Virginia Woolf: A Bibliography*. 3rd edn. Oxford: Clarendon Press, 1980.

MAJUMDAR, ROBIN, and ALLEN MCLAURIN (eds). *Virginia Woolf: The Critical Heritage*. London: Routledge & Kegan Paul Ltd, 1975. (A selection of newspaper and periodical reviews of Woolf's works, and early criticism.)

MCNEILLIE, ANDREW (ed). *The Essays of Virginia Woolf*. 6 vols. London: The Hogarth Press, from 1986 (not yet complete).

NICOLSON, NIGEL, and JOANNE TRAUTMANN (eds). *The Letters of Virginia Woolf*. 6 vols. London: Chatto and Windus, 1975–80.

SILVER, BRENDA R. *Virginia Woolf's Reading Notebooks*. Princeton: Princeton University Press, 1983.

WOOLF VIRGINIA. *The Crowded Dance of Modern Life: Selected Essays, Volume 2*. Ed.

Rachel Bowlby. Harmondsworth: Penguin, 1993. (Editor's introduction on Woolf's writing on culture and modernity.)
——*A Woman's Essays: Selected Essays, Volume 1*. Ed. Rachel Bowlby. Harmondsworth: Penguin, 1992. (Editor's introduction on Woolf's writing about women and literature and the essay as a genre.)
——*Women and Writing*. Ed. Michèle Barrett. London: Women's Press, 1979. (A collection of Woolf's essays on the subjects of the title; editor's introduction makes the case for Woolf as a socialist feminist.)

Books and articles on Woolf and her writing

ABEL, ELIZABETH. *Virginia Woolf and the Fictions of Psychoanalysis*. Chicago: University of Chicago Press, 1989. (As well as providing a reading of Woolf in relation to different stages of Freud's thinking about femininity, Abel gives an invaluable outline of the close relations between psychoanalysis and the Bloomsbury group.)
BATCHELOR, JOHN. *Virginia Woolf: The Major Novels*. Cambridge: Cambridge University Press, 1991.
BAZIN, NANCY TOPPING. *Virginia Woolf and the Androgynous Vision*. New Brunswick: Rutgers University Press, 1973.
BEER, GILLIAN. *Arguing with the Past: Essays in narrative from Woolf to Sidney*. London: Routledge, 1989. (Includes four essays on Woolf.)
——'The Body of the People in Virginia Woolf'. In Sue Roe (ed.), *Women Reading Women Writing* (Brighton: Harvester Press, 1987), pp. 83–114.
BEJA, MORRIS (ed). *Critical Essays on Virginia Woolf*. Boston: G. K. Hall, 1985.
BERNHEIMER, CHARLES. 'A Shattered Globe: Narcissism and Masochism in Virginia Woolf's Life-Writing'. In *Psychoanalysis and . . .*, ed. Richard Feldstein and Henry Sussman (New York: Routledge, 1990).
BOWLBY, RACHEL. *Virginia Woolf: Feminist Destinations*. Oxford: Basil Blackwell, 1988.
——'Who's Framing Virginia Woolf?', *Diacritics*, **21** (Summer 1991).
——'Walking, women and writing'. In Isobel Armstrong (ed.), *New Feminist Discourses* (London: Routledge, 1992).
BRAGANCE, ANNE. *Virginia Woolf ou la dame sur le piédestal*. Paris: Editions des femmes, 1984.
Bulletin of the New York Public Library, **80**, 2 (Winter 1977). (Special issue on Woolf's *The Years* and *The Pargiters*, including contributions from Mitchell Leaska and Jane Marcus.)
CARROLL, BERENICE A. '"To Crush him in our Own Country": The Political Thought of Virginia Woolf', *Feminist Studies*, **4**, 1 (1978), pp. 99–131.
CAUGHIE, PAMELA L. *Virginia Woolf and Postmodernism: Literature in Quest and Question of Itself*. Chicago: University of Illinois Press, 1991.
CAWS, MARY ANN. 'Framing, Centering, and Explicating: Virginia Woolf's Collage', *New York Literary Forum*, **10–11** (1983), pp. 51–78.
CLEMENTS, PATRICIA, and ISOBEL GRUNDY (eds). *Virginia Woolf: New Critical Essays*. Totowa, New Jersey: Vision Press, 1983.
DEFROMONT, FRANÇOISE. *Virginia Woolf: Vers la maison de lumière*. Paris: Editions des femmes, 1985. (Shaped by psychoanalytic thinking about women and literature; extract included in this volume.)
DESALVO, LOUISE. *Virginia Woolf: The Impact of Childhood Sexual Abuse on her Life and Work*. Boston: Beacon Press, 1989.

DiBattista, Maria. *Virginia Woolf's Major Novels: The Fables of Anon*. New Haven: Yale University Press, 1980.

Dick, Susan. *Virginia Woolf*. London: Edward Arnold, 1989. (A concise introduction to the novels in relation to narrative structure and experimentation.)

DuPlessis, Rachel Blau. *Writing Beyond the Ending: Narrative Strategies of Twentieth-Century Women Writers*. Bloomington: Indiana University Press, 1985.

——'Feminist Narrative in Virginia Woolf'. *Novel: A Forum on Fiction*, **21**, 2–3 (Winter–Spring, 1988), pp. 323–30.

——'Woolfenstein'. In *Breaking the Sequence: Women's Experimental Fiction*, ed. Ellen G. Friedman and Miriam Fuchs (Princeton: Princeton University Press, 1989). (On narrative technique in Woolf's *The Waves* and in Stein.)

Esch, Deborah. '"Think of a Kitchen Table": Hume, Woolf, and the Translation of Example'. In *Literature as Philosophy/Philosophy as Literature*, ed. Donald G. Marshall (Iowa City: University of Iowa Press, 1987), pp. 262–76.

Ferrer, Daniel. *Virginia Woolf and the Madness of Language*. Trans. Geoffrey Bennington and Rachel Bowlby. London: Routledge, 1990.

Fox, Alice. 'Literary Allusion as Feminist Criticism in *A Room of One's Own*', *Philological Quarterly*, **63**, 2 (Spring 1984), pp. 145–61.

——*Virginia Woolf and the Literature of the English Renaissance*. Oxford: Clarendon Press, 1990.

Freedman, Ralph (ed). *Virginia Woolf: Revaluation and Continuity*. Berkeley: University of California Press, 1980.

Froula, Christine. 'Rewriting Genesis: Gender and Culture in Twentieth-Century Texts'. *Tulsa Studies in Women's Literature*, **7**, 2 (Fall 1988), pp. 197–220.

Gilbert, Sandra M., And Susan Gubar. *No Man's Land: The Place of the Woman Writer in the Twentieth Century*. 3 vols. New Haven: Yale University Press, from 1988. (Vol. 1, *The War of the Words* (1988) and vol. 2, *Sexchanges* (1989), both include extensive discussion of Woolf.)

Gillespie, Diane Filby. *The Sisters' Arts: The Writing and Painting of Virginia Woolf and Vanessa Bell*. Syracuse: Syracuse University Press, 1988.

Ginsberg, Elaine K. and Laura Moss (eds). *Virginia Woolf: Centennial Essays*. Troy, New York: Whitston, 1983.

Guiguet, Jean (ed). *Virginia Woolf et le groupe de Bloomsbury*. Paris: Union Générale d'Editions, coll. 10/18, 1977.

Hartman, Geoffrey H. 'Virginia's Web'. In *Beyond Formalism: Literary Essays 1958–1970* (New Haven: Yale University Press, 1970), pp. 71–84.

Hawthorn, Jeremy. *Virginia Woolf's 'Mrs Dalloway': A Study in Alienation*. London: Sussex University Press, 1975. (One of the very few Marxist readings of Woolf.)

Heilbrun, Carolyn. G. *Towards Androgyny: Aspects of Male and Female in Literature*. 1964; repr. London: Victor Gollancz, 1973.

Herrmann, Anne. *The Dialogic and Difference: 'An/Other Woman' in Virginia Woolf and Christa Wolf*. New York: Columbia University Press, 1989. (Uses French feminist theory to make an original juxtaposition of Woolf with the later German writer.)

Homans, Margaret. 'Mothers and Daughters in Virginia Woolf's Victorian Novel'. In *Bearing the Word: Language and Female Experience in Nineteenth-Century Women's Writing* (Chicago: University of Chicago Press, 1986), pp. 277–88.

Hussey, Mark. *The Singing of the Real World: The Philosophy of Virginia Woolf's Fiction*. Columbus: Ohio State University Press, 1986.

Jacobus, Mary. 'The Difference of View'. 1979; repr. in *Reading Woman: Essays in Feminist Criticism* (London: Methuen, 1986), pp. 27–40.

Joplin, Patricia Klindienst. 'The Authority of Illusion: Feminism and Fascism in Virginia Woolf's *Between the Acts*'. *South Central Review*, **6**,2 (Summer 1989), pp. 88–104.

KAIVOLA, KAREN. *All Contraries Confounded: The Lyrical Fiction of Virginia Woolf, Djuna Barnes, and Marguerite Duras.* Iowa City: University of Iowa Press, 1991.

KAMUF, PEGGY. 'Penelope at Work'. In *Signature Pieces: On the Institution of Authorship* (Ithaca: Cornell University Press, 1988), pp. 145–73. (A longer version of the piece included within this volume.)

KELLEY, ALICE VAN BUREN. *The Novels of Virginia Woolf: Fact and Vision.* Chicago: University of Chicago Press, 1973.

LAURENCE, PATRICIA ONDEK. *The Reading of Silence: Virginia Woolf in the English Tradition.* Stanford: Stanford University Press, 1991.

LEASKA, MITCHELL. *The Novels of Virginia Woolf: From Beginning to End.* London: Weidenfeld and Nicolson, 1977.

LEE, HERMIONE. *The Novels of Virginia Woolf.* London: Methuen, 1977.

LIDOFF, JOAN. 'Virginia Woolf's Feminine Sentence: The Mother–Daughter World of *To the Lighthouse*', *Literature and Psychology*, **32**, 3 (1986), pp. 43–59.

LILIENFIELD, JANE. '"The Deceptiveness of Beauty": Mother Love and Mother Hate in *To the Lighthouse*'. *Twentieth-Century Literature*, **23**, 3 (October, 1977), pp. 345–76.

LONDON, BETTE. *The Appropriated Voice: Narrative Authority in Conrad, Forster, and Woolf.* Ann Arbor: University of Michigan Press, 1991.

MARCUS, JANE. *Virginia Woolf and the Languages of Patriarchy.* Bloomington: Indiana University Press, 1987.

——*Art and Anger: Reading Like a Woman.* Columbus: Ohio State University Press, 1988.

——(ed.). *New Feminist Essays on Virginia Woolf.* London: Macmillan, 1981.

——(ed.). *Virginia Woolf: A Feminist Slant.* Lincoln: University of Nebraska Press, 1983.

——(ed.). *Virginia Woolf and Bloomsbury: A Centenary Celebration.* London: Macmillan, 1987.

MARDER, HERBERT. *Feminism and Art: A Study of Virginia Woolf.* Chicago: University of Chicago Press, 1968.

McLAURIN, ALLEN. *Virginia Woolf: The Echoes Enslaved.* Cambridge: Cambridge University Press, 1973.

MEISEL, PERRY. *The Absent Father: Virginia Woolf and Walter Pater.* New Haven: Yale University Press, 1980.

MEPHAM, JOHN. 'Figures of desire: Narration and fiction in *To the Lighthouse*'. In Gabriel Josipovici (ed.), *The Modern English Novel* (London: Open Books, 1976), pp. 176–85.

——'Mourning and Modernism'. In Clements and Grundy, pp. 137–56.

MILLER, C. RUTH. *Virginia Woolf: The Frames of Art and Life.* New York: St Martin's Press, 1989.

MILLER, J. HILLIS. *Fiction and Repetition: Seven English Novels* Cambridge: Harvard University Press, 1982. (Concludes with deconstructive essays on *Mrs Dalloway* and *Between the Acts*.)

MINOW-PINKNEY, MAKIKO. *Virginia Woolf and the Problem of the Subject.* Brighton: Harvester Press, 1987. (A reading that uses the work of the French psychoanalyst and semiotician Julia Kristeva.)

MOI, TORIL. 'Introduction: Who's Afraid of Virginia Woolf?' In *Sexual/Textual Politics: Feminist Literary Criticism* (London: Methuen, New Accents series, 1985), pp. 1–18.

MOORE, MADELINE. *The Short Season between Two Silences: The Mystical and the Political in the Novels of Virginia Woolf.* Boston: Allen and Unwin, 1984.

NAREMORE, JAMES. *The World Without a Self: Virginia Woolf and the Novel.* New Haven: Yale University Press, 1973.

POOLE, ROGER. *The Unknown Virginia Woolf.* Cambridge: Cambridge University Press, 1978. (While leaving in place the categories of madness and sanity, Poole inverts the representation of Woolf as mad: instead, it is her husband and a patriarchal psychiatric establishment.)

RICHTER, HARVENA. *Virginia Woolf: The Inward Voyage*. Princeton: Princeton University Press, 1973.

ROE, SUE. *Writing and Gender: Virginia Woolf's Writing Practice*. Hemel Hempstead: Harvester Wheatsheaf, 1990.

ROSE, PHYLLIS. *Woman of Letters: A Life of Virginia Woolf*. London: Routledge & Kegan Paul, 1978.

ROSENMAN, ELLEN BAYUK. *The Invisible Presence: Virginia Woolf and the Mother–Daughter Relationship*. Baton Rouge: Louisiana State University Press, 1986.

——'Sexual Identity and *A Room of One's Own*: "Secret Economies" in Virginia Woolf's Feminist Discourse'. *Signs*, **14**,3 (Spring, 1989), pp. 634–50.

RUOTOLO, LUCIO P. *The Interrupted Moment: A View of Virginia Woolf's Novels*. Stanford: Stanford University Press, 1986.

SCHLACK, BEVERLY ANN. *Continuing Presences: Virginia Woolf's Use of Literary Allusion*. Pittsburgh: Pennsylvania State University Press, 1979.

SHOWALTER, ELAINE. 'Virginia Woolf and the Flight into Androgyny'. In *A Literature of their Own: British Women Novelists from Bronte to Lessing* (1978; repr. London: Virago, 1979), pp. 263–97. (One of the rare feminist readings to be negative about Woolf's writing.)

SPIVAK, GAYATRI CHAKRAVORTY. 'Making and Unmaking in *To the Lighthouse*'. In *In Other Worlds: Essays in Cultural Politics* (New York: Methuen, 1987), pp. 30–45.

SQUIER, SUSAN. *Virginia Woolf and London: The Sexual Politics of the City*. Chapel Hill: University Press of North Carolina, 1985.

TAMBLING, JEREMY. 'Repression in *Mrs Dalloway*'s London'. *Essays in Criticism*, **39**, 2 (April 1989), pp. 137–55.

WAUGH, PATRICIA. *Feminine Fictions: Revisiting the Postmodern*. London: Routledge, 1989. (Includes chapter on Woolf as more than a modernist.)

ZWERDLING, ALEX. *Virginia Woolf and the Real World*. Berkeley: University of California Press, 1986.

Index